T0177126

Scripts, Grooves, and Writing Machines

Scripts, Grooves, and Writing Machines

Representing Technology
in the Edison Era

Lisa Gitelman

Stanford University Press, Stanford, California

Stanford University Press
Stanford, California

© 1999 by the Board of Trustees of the
Leland Stanford Junior University

Printed and bound by CPI Group (UK) Ltd,
Croydon, CR0 4YY

CIP data appear at the end of the book

IN MEMORY OF MY DAD

Howard M. Gitelman

Acknowledgments

This book would not have come about without the gracious and able assistance of many readers and the support of many friends. First among them, I would like to thank my colleagues, past and present, at Rutgers University for offering an initial line of defense against my most half-baked schemes and inarticulate enthusiasms. Theresa Collins and Gregory B. Field proved themselves particularly generous, always asking hard questions, suggesting good sources, and proving the merits of true collegiality. I would also like to thank Esther Katz, Miranda Paton, and the reviewers at Stanford University Press for their intelligent, exacting readings of the manuscript. Along the way, Leslie Fishbein, Thomas Y. Levin, and Ron Thomas made valuable suggestions and offered words of support, as did Michael Kucher, Chuck Bazerman, Jerry Fabris, Bob Rosenberg, and Paul Israel. Extra appreciation is due Pat Crain, plus the gals from Columbia, for their insight, comradeship, and commiseration, and also Hill, Alix, and Claudia, who, along with many friends outside of academe, kept reminding me that they wanted to read what I was writing (and when was it going to be published, anyway). While working on this project I enjoyed Carl Kaestle's NEH summer seminar for college teachers, Meredith McGill and Joan Shelley Rubin's American Antiquarian Society seminar on the history of the book, as well as the Department of American Studies' own faculty colloquia at Rutgers. To the sponsors and organizers of, and many participants in, such enlivening endeavors, thank you. I must also thank the Edison National Historic Site in West Orange, New Jersey, where I have enjoyed using the archives under the auspices of the Thomas A. Edison Papers Project. Part of Chapter 3 appeared first in *The Musical Quarterly* 81 (1997) and is republished here with the permission of Oxford University Press. All illustrations, except as otherwise noted, appear courtesy of the Edison National Historic Site, U.S. Department of the Interior, National Park Service.

L.G.

Contents

Illustrations

Illustrations

Scripts, Grooves, and Writing Machines

Introduction:
Writing Things Down,
Storing Them Up

This book is about machines for writing and reading in late-nineteenth-century America. Its purpose is to explore writing and reading as culturally and historically contingent experiences and, at the same time, to broaden the current widely held view of technology in its relation to textuality. The main character in this narrative is the phonograph, invented by Thomas Edison in 1877. That I make the phonograph my protagonist plays upon the gist of my argument: isolating and centering machines in this way, to essentialize them as *the* phonograph or *the* computer, is misleading and denies their history. Edison identified his phonograph as a textual device, primarily for taking dictation. With this mandate, the invention emerged from Edison's laboratory into and amid a cluster of mutually defining literacy practices, texts, and technologies, among them shorthand reporting, typescripts, printing telegraphs, and silent motion pictures. Even Edison's own famous light bulb, now a universal icon for "I have an idea," had to make sense within an ambient climate of textual and other representational practices, a climate it would, in fact, have an ample share in modifying.

This inquiry, in part, represents my reaction to studies of machines for writing and reading at the end of the twentieth century. Such accounts consistently draw a measure of their authority from comparisons between the past and present and, in particular, between the printing press and computer as agents of change. This shift from Gutenberg to hypertext has been greeted with celebration by some, sackcloth and ashes by others, while the emotional tenor of response tends to deflect attention away from questions about shifting per se. The most schematic accounts simply jump from the logic of print in the sixteenth century to a new logic for digital communications in the twenty-first, as if five hundred years had not happened, or as if Caxton and Carlyle, Paine and Pound, could have experienced textuality in the same way. Even the more extended narratives of George Landow, David Jay Bolter, and Richard Lanham, so rich in detail about the new world order, elide crucial developments toward the end of the nineteenth century that together prefigure most of the "revolutionary" aspects of digital, hypertextual networks. The pending ubiquity of multimedia, of paperless offices and personalized newspapers, as well as the supposed democratization of information and liberating proliferation of "virtual" identities were also imagined in association with predigital technology. Here, I seek to interject a corrective portion of the missing detail; this work's grounding contention is that these same accounts generally distort the character and contexts of literacy and textuality in modern life. In particular, they fail to explore technology as plural, decentered, indeterminate, as the reciprocal product of textual practices, rather than just a causal agent of change.

My focus is on experiments and innovations in the area of *inscription*. In the manner of German theorist Friedrich Kittler, I see mechanized inscription as integral (though certainly not unique) to the climate of representation that emerged toward the end of the nineteenth century and has dominated the twentieth. Its features are quickly mapped: Edison stumbled across the idea of mechanically inscribed sound during his work improving Alexander Graham Bell's telephone. He jotted hastily in his experimental notebook, "There[']s no doubt that I shall be able to store up & reproduce automatically at any future time the human voice perfectly." Even if his accomplished phonograph was many sleepless nights in the future, Edison self-confidently set his sights on a new form of inscription. His single, jumbled sentence identifies the fundamental properties of inscribed communication that distinguish it from communication generally. Telephones reproduce speech at a distance, but phonographs both reproduce and conserve, able to reproduce again and again "at any future time," because of the delicate spirals inscribed on the surface of records. The same parameters of economy and durability that characterize "storing up" sound for later mechanical reproduction helped animate such contemporary phenomena as the tensile bureaucracy of managerial capital, the ideal of objectivity in the professions and media, and the success of new popular culture forms. Economy and durability informed new modes of inscriptive duplication, such as the office mimeograph, which allowed bureaucrats to have their copy and send one too. Likewise, economy and durability characterized considerations of photography and then motion pictures, which stored up sights and movements. Incidents as fleeting as the pulsations of the heart and activities as evanescent as the private use of electrical current were captured, registered, metered, and read in new mechanical ways. Some of these inscriptions were more transparent as representations than others; some were more textual, some more graphic. Many, like the grooved surfaces of phonograph records, provoked explicit questions about textuality, about how some inscriptions might or might not be like texts.

Inscription is a form of intervention, into which new machinery continues to interpose. Ink is imposed on paper, while pens and keyboards intrude into the posture of hands. Grooves are incised into phonograph records, while sound echoes in our ears. The genealogies of inscription allow what anthropologist Michael Taussig calls "particular" histories of the senses, as different media and varied forms, genres, and styles of representation act as brokers among accultured practices of seeing, hearing, speaking, and writing. There is what Jonathan Goldberg calls a "history of technology that is also the history of 'man,' the programmed/programming machine: the human written" (24). This is the kind of history that *Scripts, Grooves, and Writing Machines* seeks to locate. However, I have explored its contours and implications within a cultural history of experience, rather than the anti-interpretive, or "posthermeneutic," philosophy of poststructuralists such as Kittler, where technology tends to retain a symbolic character.[1] My discussion offers a less determined, less determinist version of technology, in part by locating writing machines and other textual devices in the instances of invention, in narratives that show each machine, device, or process to have been authored and appropriated out of many different possibilities relevant to the making of meaning. I begin with the idea that inventing new ways to write or new kinds of writing presupposes a model of what writing and reading are and can be. If the model is too eccentric, then the invention may not work, or the model might suit some relevant social groups and not others. If the model is negotiable within or against existing models, then the invention has a chance of appropriation and dissemination. In this way, shorthand alphabets, phonographs, typewriters, and other nineteenth-century innovations in the area of inscriptive practice are so many theories of language and textuality. They are not *the* theory of language held by all Americans at the time; they are not "our" theory of language. Instead, they are modest, local, and often competitive embodiments of the way people wrote, read, and interacted over the perceived characteristics of writing and reading.

This perspective has a couple of advantages. First, it generalizes from consensus and thus from technologies that eventually proved unworkable as well as workable, since both are revealing sites of negotiation. In this sense the view has "symmetry" according to the sociology of knowledge: it hints at a more complete compass of experience, at the category of technical "workability," in the same way that noncanonical and "sub"-literary works and everyday textuality hint at the category of "literariness" in any epoch. Thomas Edison's "electric pen" stenciling devices, to take one example, proved far less successful than office mimeographs as a means of textual duplication, as did numerous other contrivances that never made it to market at all. Yet electric pens and the rest of the objects in the discard pile are no less worthy of study than mimeographs. Like mimeographs, electric pens can be located among an inventor's, promoter's, and consumer's preconceptions and receptions, each of them pliable. Together they permit an interrogation of textual duplication as a socioeconomic and a linguistic fact, part of emerging business practices, shifting labor cohorts, and the burgeoning potential of writing to be all over and all the same, somewhere between manuscript and print. The eventual failure of electric pens reveals something; the eventual success of mimeographs probably tells a little more; looking into the social negotiations of failure and success promises to reveal the most.

The second advantage of considering machines for reading and writing as consensual, embodied theories of language is the potential the approach offers for an additionally symmetric account of cultural production and consumption. It does no good to look at theories of language foisted on a blank public by individual and frequently idiosyncratic inventors. No inventor is the beginning of a circuit, sprung whole, like Athena from the head of Zeus. No public is a blank receptor. And "foisting" is far too simple a verb for what I want to call negotiation and appropriation. Like text, new technology is not objectively consumed. As Roger Chartier observes of the former, "Experience shows that reading is not simply submission to textual machinery" (156). Nor, as Marshall McLuhan and Jean Baudrillard are both famous for suggesting, is the experience of text simply a submission to the inscriptive medium of print or computing, the mechanical modes or electronic codes of its presentation and reproduction.

The dual symmetries of success and failure and of producer and consumer appeal in theory but prove difficult in practice, because success and production form such powerful forces of historiographic orientation. Technologies that succeed exert a teleological tug: mimeographs tend to erase electric pens on the way to photocopiers and facsimile machines. The physical and commercial shape of the pens is excised from memory and so is the partially linguistic phenomenology of their use. One forgets what the pens meant by forgetting *how* they meant. And then this amnesia is itself forgotten, in direct relation to the seriality, the Progress, unthinkingly ascribed to present and future inventions. As moderns and as consumers, we have been conditioned to think that technologies supersede each other one by one, the present ever liberating us from the past. Added to the double problem of forgetfulness, the producers of technology always leave more traces of themselves than consumers do. The makers and purveyors of mimeographs proportionally erase the many more numerous users of their machine by dominating the historical record. Consumers of failed electric pens, by this account, sit smack in a blind spot, equally obscure to history and discomfiting to historiography. Their technology did not "win." They were "only" its users. Even when the fewer users of the pen can be identified—Charles Dodgson, a.k.a. Lewis Carroll, had one[2]—it is difficult to renovate their experiences from the level of anecdote or emblem to the level of evidence. I am describing a challenge, not an impasse: being careful with questions and patient at archives can unearth a variety of sources for analysis that can help cut across teleological habits. Technology, whether inscriptive or not, involves a lot of paper. Machines get some of their meaning from what is written about them in different ways and at specific junctures, in research plans, patent applications, promotional puff, and so on. Writing machines, in particular, get some of their meaning from the way they are used, including the writings they produce.

If paperwork can reveal so much about technology, then technology, like science, has a rhetoric of its own. It relies upon rhetorical processes, the conventions of which contribute to a "thick" description of culture, revealing the way American culture sees itself and hinting at the way it identifies and legitimates "the facts." This follows from Edwin Layton's observation, unremarkable among historians of technology today, that technology constitutes a form of knowledge. Rather than an inert and hermetic materialism, technology presupposes a "spectrum," according to Layton, ranging from an idea, through a design, and finally to an artifact and its apprehension and use. Each element in this spectrum involves a different and varied knowledge practice, combining the nonverbal and the verbal, know-how and knowing, *techne* and *logos*. Mechanical, material components and systems are only the most familiar expressions of technology as a form of knowledge. They very rarely exist without linguistic and graphic compliments, labels, descriptions, drawings, and diagrams. In this light technological innovation becomes a process of selecting, sifting, and circulating messages, from the proverbial drawing board to the marketplace and then the drawing room. Artifacts become knowable in part because they are enmeshed within the back and forth and round about of telling what they are, and because telling devolves upon discernable rhetorical conventions, like genres and specialized vocabularies, that are themselves largely the result of unconscious consensus. Economic realities tend to enforce this rhetorical character of technological knowledge by requiring the literature of patents and the literature of commercial promotion. Both the need to identify property and the desire to exchange it ensure an insistently rhetorical character almost unmatched in science, where disciplinary pressures stand in place of commercial ones.

Not only does the rhetorical character of technology allow for a "softer" determinism by which machines are not simple, unitary influences on writing but also the same character permits a degree of critique that has eluded all but a few humanist (and those particularly feminist) attentions to science. If technology is a form of knowledge, then it can be conflicted with doubt and contradiction, with assumptions and anxieties, just like other forms of knowledge. The answer to Langdon Winner's provocative question, "Do artifacts have politics?" is affirmative, if only most obviously in cases like the gender politics of Dalkon Shields, or the racial politics of Pullman sleeping cars. Fountain pens and typewriters can be just as ideological, just as much superstructure as infrastructure. Culture insinuates itself within technology at the same time that technology infiltrates culture. Rhetorical analysis provides one way to glimpse the localities of both insinuation and infiltration while testing the usefulness of their directional semantics. It also tests the metaphorical conventions resident in shaky, long-lived assumptions that technological innovation and writing are uninvolved with one another, or that causality is a one-way street along which technology drives history.

This underlying sense that technology is enmeshed within textuality, that machines are discursively and physically constructed, is a view garnering surprisingly little direct attention. Historians of technology have only lately begun to ponder what they call the "interpretive flexibility" of artifacts possessing "technical content" for "relevant social groups." Drawing upon earlier work in the sociology of scientific knowledge these scholars have tried to set aside the habitual opposition that both divides and defines their discipline, the one between internalist and externalist histories. Internalists practice a sort of formalism, attending more narrowly to how things work, the way one telegraph instrument adapted the form or function of another. Externalists, by contrast, locate things more amid political, economic, and cultural contexts. The newer school of social constructionists rejects both thing and context as separate or separable units of analysis.[3] According to this view, an invention succeeds not because "it works," but rather it is described as "working" because it succeeds amid prevailing and possibly competitive expectations. Technological function remains something to explain; it does not comprise an explanation in (or of) itself. To put this another way, artifacts are themselves astute, yet they cannot answer *all* of our questions about why one invention becomes accepted and another does not, any more than a novel, for instance, can answer all of our questions about how it was written and how it was read. Artifacts cannot even answer with sufficient precision why one model of a machine is "better" than another, just as an edition of poetry cannot speak completely to its own superiority or popularity over other editions.

Despite my analogy, the discursiveness of technology has sometimes been hard for scholars of literature, linguistics, and communication to rehearse. Even the most committed social-constructionists seem at times to fall back upon technology as a stable ground amid the roiling, discursive sea. So Jonathan Boyarin, for instance, chides

that reading is not a technology, in his superb collection of ethnog-
raphies on the subject. And Brian Street insists that literacy is not a
technology, when he rightly critiques the determinism of scholars
pronouncing the cognitive rewards purchased by literacy over and
against orality. In both cases and with the best of intentions, technology is a determinist standard, an unchanging and surreptitiously
ideological norm that the constructionist project perpetuates in order to delimit its own comparative critique.[4] The literary scholars
who have articulated more nuanced versions of technology, broadly
speaking, are those who experience texts *as* technologies in the work
of textual editing and bibliography. Editors and bibliographers, such
as Jerome McGann and D. F. McKenzie, are accustomed to the idea
that texts are material things, the very design and physicality of
which are important to their meanings as texts. As McGann puts it,
"The body of the text is not exclusively linguistic" (13). It is a
"laced network" of linguistics and bibliography. And bibliographic
fact encompasses the physical artifact, acknowledging its production, circulation, and consumption as such. This is a textual materialism that enrolls print and what McKenzie calls "non-print texts"
within the interests of material culture studies without ceding the
special status of texts as authored, literary, or otherwise unique
commodities. The loose, disciplinary designation for much textual
materialism has lately been, in English, "the history of the book,"
though *the* book is less than adequate as a term for subjects as broad
as nonprint texts and "alternative information technologies."[5]

Particularly suggestive for my purposes have been recent studies,
such as those by Richard John and Michael Warner, that locate the
social meanings of texts within the early American republic. John's
history of the post office treats texts in their physical circulation and
imagined circulability among readers. He demonstrates persuasively
that communications networks have to be bureaucratic before they
can be technological. Warner renders the "cultural meaning of print-
edness" as an instrument and indicator of social transformation
within which the "force" of print technology and the act of reading
together became the engine of a bourgeois public sphere in America

(xi). Both scholars have been influenced by the work of Jürgen Habermas, who plots the origin and structural transformation of the public sphere in Europe as the creation and then corruption of an abstract level of rational, critical discourse among bourgeoisie. This modern public sphere was created in some degree by a new subjectivity of print in the late seventeenth century—by reading in a new way and by reading novels—and corrupted in the eventual passivity of consumer culture.[6] Like Benedict Anderson and Angel Rama, Habermas grants a great deal of weight to the circulation of print as socially constitutive and transformative. This is a particularly rich background with which to explore the related matter of inscription in the late nineteenth century.

The study of inscriptions shows the realm of writing and reading, of symbolic action and experience, in its proximity to objects and machines. From ancient marks on clay or carvings in stone to the printed labels affixed to commercial goods today, inscriptions insistently belie their own double character, both material and semiotic. Modern technology has made some features of this doubleness seem particularly arcane. For example, the original electric meters of the 1880s were really halves of little batteries; to "read the meter" a technician had to remove a zinc electrode and weigh it in order to determine the amount of ion deposit, the amount of electrolytic action, which in turn indicated the amount of current that had passed through the system. Those ion deposits, like the ion deposits on photographic plates or strips of celluloid, are the stuff of inscription. They are the double-sided boundary at which the built system both represents technology and is technologically represented. With the same doubleness, the word *Representing* in my title is both verb and adjective, looking toward technologies represented as well as toward representations and inscriptions generated technologically by the typewriter, the phonograph, and the like. It is a profitable doubleness and, I will argue, particularly revealing of modernity and modern subjectivity.

As if it were curling up with a good book in a halo of light, the electric company wants to keep reading meters, so the rest of us will keep paying bills. New inscriptions like these have regulated modern experience, making life more legible in complicated, public ways, signaling changes to the context and thereby the complexion of writing and reading. And changes to writing and reading matter in large measure because they equal changes to writers and readers. New inscriptions signal new subjectivities. Racial, national, sexual, and other properties of selfhood seemed newly marked and conflicted by mimesis and difference, as missionaries used shorthand to "take down" the languages of indigenous peoples, as later travelers would "take" phonograph records and films. Modern inscriptions appeared freshly laden with meaning, forensically hinting at normative categories of legality, health, reason, whiteness, and so forth. Starting in 1891 British readers of the *Strand* saw how Sherlock Holmes could trace authorship in the unique imprint of an individual typewriter. France treated the world to Bertillon's bizarre graphology, his theory of handwritten "self-forgery," which helped convict Alfred Dreyfus in 1894.[7] In 1896 Roentgen's discovery of X rays caused a sensation, opening the human body to view like the pages of a book.[8] And in 1898 the newly yellow press helped inflame the tempers of American imperialism. Phonograph and music roll companies meanwhile cashed in on the popularity of "coon" songs, while Edison's "kinetograph" studio practiced filming Italian organ grinders and their monkeys from New Jersey, before sending celluloid versions of Buffalo Bill's Wild West Show out to urban "kinetoscope" exhibitors.[9]

New inscriptive forms interrupted and helped ratify, stretch, or commodify contemporary parameters of identity; they equally intervened between private and public life. Publishing, that is, making public or evident, was a cardinal rationale behind storing up and writing down. Even within the relative privacy of homes, business offices, and medical clinics, new kinds of authorship or new kinds of reading helped warrant a new sense of public existence. Some modes of inscription, such as the shorthand alphabets I describe in Chapter 1, were used as part of a self-conscious construction of the public weal. Some, like X-ray prints, were made public by subverting the private, literally getting under its skin. Others, like the typescripts I address in Chapter 5, succeeded amid relevant shifts in socioeconomic geography: the world of business offices became more public precisely in becoming more gendered. Together with the new inscriptive forms of popular amusement, inscriptions like these helped imagine a public sphere radically different than the one Habermas renders for the late seventeenth and eighteenth centuries. A neutral, impersonal public arena for the prospect of public knowledge continued in the ongoing formulation of the newer "human" sciences, described by Michel Foucault, as well as in the ideal of objectivity within the professions and the ever-deepening trust in quantifiable bases for public policy, described by Michael Schudson and Theodore Porter, respectively. But at the same time, "going out" to phonograph and kinetoscope parlors relied upon a changing version of "out," upon changing experiences of public knowledge, public space, self, and community. By the late nineteenth century, nonprint, inscriptive media had become rich and vigorous "allies" to print in the construction of imagined communities,[10] joined so obviously in the twentieth century by noninscriptive forms, such as the rather late in coming sociality of the telephone ("touch someone"; see Fischer), broadcast radio ("the listening audience"), and then television ("viewers like you"). The changing public sphere was evident in the phonograph and kinetoscope parlors, then in the nickelodeons, where crowds sat silently together in the dark, as well as in the proliferation of mass media, with its defining consumption of public taste in the form of fads, hits, bestsellers, and stars. It was evident as well in the imagination of technology as an abstract, shared public good, in the so-called "public" utilities, for instance (generally big private companies that carved up American markets under the glassy eye of municipal authority), or in the eventual imagination of technology as a shared public menace to labor, to the environment, to health, or to peace.

Narratives about phonographs prove one way to add a little flesh to these schematics about a sense of modern, public existence and correspondingly modern subjectivities, because the early identification of the machines as textual devices was so decisively corrected in their popularity as musical instruments. Phonographs were the children of Anderson's "print capitalism," adopted and reared by mass culture. They were introduced as objective instruments of public knowledge and were appropriated as amusing media of public taste. Neither histories of information control, like James Beniger's *Control Revolution*, or histories of public amusement, like David Nasaw's *Going Out*, can get at both sides of the equation. Issues of objectivity and mimetic power that attended the phonograph as a textual device pursued it with vigilance into popular culture. How could the same instrument of rational, critical discourse become the object of passive consumption? How could "consumption" in cases like these become as passive as Habermas and the bleaker accounts of the Frankfurt School critics suggest? Narratives about phonographs open questions like these, while they hint at just how deeply misserved the public is by assuming, as so readily happens today, that print and nonprint media are antithetical sources of meaning in the modern context. Print culture and nonprint media evolve in mutual inextricability. The phonograph and contemporary inscriptive forms were deeply dependent upon reworkings of the social and economic relations of textuality, of print culture and print capitalism. They engaged literacy practices in toto, the cognitive and the somatic, the semiotic and the social. They helped question authors and readers as subjects and modify the experienced subjectivities of speakers, performers, publishers, and literates. In doing so they kept intervening into dynamic constructions of private and public, community and difference.

Each of the five chapters that follow begins with a different version of the same scene of writing: Edison's invention of the phonograph, and the machine's own traverse from private to public. This is a "scene of writing" in the way that poststructuralists like so much, a self-reflexive authorial vignette within which a phonograph stylus, Edison, and his laboratory all make sense of making sense, and do so in revealing ways. Edison's invention articulated himself as an inventor and articulated his phonograph as both textual and mechanical. From this same scene, each chapter heads in a different

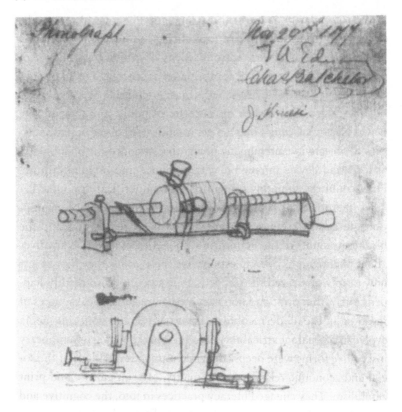

FIGURE 1. The phonograph invented. Edison laboratory drawing, November 1877.

direction: Chapter 1 backgrounds; Chapter 2 imagines; Chapter 3 authorizes; Chapter 4 labels; and Chapter 5 supplements. Together the five vectors describe a genealogy for phonographs and other inscriptive technologies as machines for writing and reading. Together they encounter contemporary experiences of authorship and reading, the discomforts that the varied and questionable textuality of new inscribed forms seemed to inspire, and the largely uncalculated negotiations that helped those inscriptions make sense in a changing world. The negotiating table was set with powerful assumptions, among them the dichotomous oppositions of ear and eye, mouth and page, private and public, experience and evidence, "man" and machine. These dualisms operated in tandem and in opposition in the appropriation of a sense of textuality somehow suited to the modern moment. They can be read in the five genres that orient respective chapters: the shorthand manuals, "idea letters," patents, product labels, and "automatic writings." This is not a history of the phonograph or a story about Thomas Edison. Plenty of both exist.[11] Instead, this is an interdisciplinary method, a bibliography for machines.

Chapter 1 backtracks in order to explain the context within which the phonograph seemed possible, desirable, and textual. These are the cultural preconditions for relevant technological change. It is only with blurry and uncritical hindsight that Edison's two models look like the preexisting music box and the human memory.[12] Instead, Edison's primary model was shorthand, while his secondary models were versions of the telegraph, telephone, and phonautograph, a scientific instrument for drawing sound waves. This chapter outlines the history of shorthand in the nineteenth century in order to show the issues of textuality to which the phonograph proved an a posteriori response. Shorthand was a technological application of the inchoate "science" of phonetics, expressive of deep-seated assumptions about language and authority. The inventors of shorthand, men who competitively and vociferously promulgated their own shorthand alphabets, and the consumers of shorthand, the reporters who "took" and inscribed government activities, all found themselves multiply testing the bounds of textuality. They rattled the category of authorship, worried at the limitations of their own bodies as inscriptive tools, and dreamed of navigating a course between the legible *representation* of aural experience and some more perfect, legible *reproduction* of the same thing.

Chapter 2 imagines the future of the phonograph. I begin with the preconceptions of Edison and his associates at the moment of its invention. Their preconceptions turned out to be *mis*conceptions about the stenographic function of the machine and the textual nature of its mechanically "written" inscriptions and mechanically "read" reproductions. Mistakes like these can be very revealing, both in their misapprehension of technology and in their misapprehension of textuality. After presenting readings of two speculative texts by the inventor, I rely primarily upon a group of letters written to Edison, in which the authors propose their own inventions. Thousands of Americans seized upon technological innovation as an act of self-determination, upon technology as a private experience of public life. A veritable cult of inventing, with Edison as its icon, promised that having a good idea and "going public" was a path to success. The letters of these amateurs survive in the Edison archive. Together they offer an important register of the experienced role technology possessed in American lives. They second Edison's interests in inscriptive technology at the same time that they reveal the patterns of expression with which technological innovation was broadly characterized and the psychical baggage with which it was frequently laden. Their proposals of different adaptations of the phonograph and other devices demonstrate assumptions about the character of language and textuality. They show that, despite a shared sense of inevitability that technological change would affect literacy practices in some way, individual experiences of text and technology were extremely particular, personal, and often contradictory. Their recovery forms an important hedge against the problematic customs of generalizing "our" interests in technological change and totalizing the ways "we" have apprehended and use various machines.

Chapter 3 locates the phonograph amid the authorizing textual practices of the U.S. Patent Office and against the authorizing strategies of American copyright law. Patents are the official textual identity of technology, and their conventions allow a glimpse at deepseated assumptions about technological knowledge that have been federally legitimated. The rhetoric of patents assumes a sufficiency and transparency of language, that inventions can be completely and objectively described, while the process of patent law continually redefines that sufficiency and transparency in the courtroom, ever enforcing new interpretations in individual judgments on infringement complaints. The sufficiency and transparency of language proved doubly problematic at the beginning of the twentieth century, because the hectic commercial exploitation of new inscriptive forms forced law and government to decide the textual properties of inscription and to legislate both what authorship had to do with mechanical reproduction and what reading had to do with machines. In forging a constitutionally and economically acceptable musical copyright provision for the Copyright Act of 1909, Congress admitted that machines could read. Machine-readable text was hardly a new phenomenon—Jacquard looms "read" punch cards in the first years of the nineteenth century—but now the textuality of records and music rolls was commonplace and federally affirmed. Affirmation seemed possible amid the technological artifacts, judicial decisions, and business structures of the day, but it was further enabled by much broader and more diffuse aspects of cultural change. The new copyright act displaced the visuality of reading and acknowledged the problematic visuality of recorded performances: you didn't have to look to read, and you couldn't see the stage. In doing so it coincided with displacements of other kinds, most notably the displaced visuality of racial identity in contemporary America. Immensely popular "coon" records of the day snatched the racial identity of performers from view, while they preyed upon the partly visual conventions of minstrelsy, upon haunting after-images of blackface. And the Supreme Court's infamous *Plessy vs. Ferguson* decision rejected visuality as relevant to race at all, in the circumstances of its "separate but equal" decree of 1896. Invisible coon-song performers were supposed to "sound 'black,'" according to long-lived, raucous, and racist norms of seeing, sounding, and blackness, while "white" or "light"-skinned Homer Plessey simply *was* "black," by dint of an invisible and essential racial difference. American experiences of difference as complex and troubling as these, whether of racial, ethnic, class, or national identity, were unadmitted legislators in the debates over musical copyright and machine-readable text.

Chapter 4 explains the technology transfer that takes place in the conversion of an invention into a product. Its title is "Paperwork and

Performance," because these are twin pressures bearing forcibly and immediately on inventions as potential products. The performance of inventors becomes accomplished with the performance of their inventions, discernable first in the speculative paperwork of the laboratory and then in the speculative paperwork of market placement. Paperwork and performance resonate with particular richness when the invented products in question are cultural forms, like phonograph records and motion picture films. Mechanical reproductions like these are themselves performative and inscriptive. Experiences of them change according to available categories of representation. The first scratchy phonograph records, for instance, reportedly sounded "just like" the sounds they recorded, while the perceptual condition of sounding "just like" has continued to change over time and according to expectation and technology, most recently from the standards of analog to the standards of digital recording. Tracing changes to the formal conventions of product labels is one way to calibrate the experienced ontology of a product or products. Labels narrate the intended meaning of a product, its recognizably salient features as produced, saleable, and consumable. So the changing conventions of early phonograph record labels and film titles provide access to the experiences of hearing and seeing these cultural forms as emergent articles of trade and mass consumption at the end of the nineteenth century. The form and content of each new label indicates a new understanding of the representational product. Successive adjustments to the labeling process further reveal successive reorientations within the category and status of representation as such, as the producers of phonograph records and films struggled to capitalize on so-called "real life" motion pictures and "live" recordings.

Chapter 5 adds the complimentary textuality of typewriters to the experienced textuality of phonographs. Reading and writing were interrogated by both machines during the same period, though typewriters and typescripts intervened more aggressively into the experiences of writing. If phonographs became everyday "reading" machines, typewriters were even more certainly the writing machines of modern life. A predictable myth of supersession arose, within which typewriters would soon banish pens and pencils,[13] while the inter-session of modern machines between author and page helped adapt the subjectivities of authorship and publication in new ways, in part by engaging the same categories of aurality, orality, and textuality that phonographs also engaged. While phonographs mystified cultural production and performance in "live" recordings, typewriters generated a different kind of "noise," a static or supplement that arose as part of the inscriptions they produced. What interests me here is this aspect of noise, which I defer in my earlier discussion of shorthand. It is the clatter of intention, inspiration, and involuntarism that attends both writing and authorship as human knowledge practices and that seems to have become differently charged with gender and sexuality around the turn of the century. Typists and typing made sense in ways that remain relevant today, even though so many now "type" on the keyboards of word-processing computers. They came to make sense partly by recycling the social meanings of writers and writings of a very different sort—the automatic writers and writings of American spiritualism. During the 1890s "automatic writing" was a phrase that applied doubly to the work done on typewriters and during seances, by secretaries and mediums, both of whom were usually women. Exploring this coincidence of meanings serves to unpack McLuhan's terms "medium" and "message," to historicize "the Media," at the same time that it suggests ways in which new experiences of textuality are adapted within the social circumstances they help to transform.

The following pages draw comparisons between the textual machines of 1877–1914 and the textual machines of today when those comparisons seem relevant and helpful. Because machines for writing and reading are the partial embodiments of historically and culturally contingent experiences of textuality, many historical comparisons can be points of refraction, if not obscurity. Yet comparisons are always tempting. A coda on the "(hyper)textuality of everyday life" suggests ways in which recent accounts of digital textuality rely upon historically comparative explanations. The coda asks explicitly why so many recent accounts of hypertext ignore the period from

1877 to 1914, a span many cultural critics and political, social, and cultural historians have long identified as crucial to modernity and to the cultural axis of modernism/postmodernism. I suspect this oversight arises from the same intellectual reservoir as the oversight of recent historians of computers and computing, who generally do not mention, or who dismiss as "less interesting," the linguistic and textual functions of the digital machines designed first for calculation and then for data processing.[14] That is, I think both oversights arise in the conflicted status of texts as cultural artifacts, particularly in their relation to literary works as separable and privileged things, in a cultural and political climate or tradition that chronically grants too much agency to technology. Technology—I agree with the Marxist critic Fredric Jameson—has a "purely local validity" as an explanation of culture (25). Yet for both the critics who bracket hypertexts as transtextual accomplishments and the historians who bracket computers as paratextual devices, technology generally possesses a metahistorical significance. It is either the proactive consummation of poststructuralist literary theory, or it is the proactive consummation of numeracy and bureaucracy. *Scripts, Grooves, and Writing Machines* offers instead the mutual relations of technology and textuality.

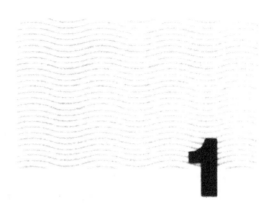

Making History,
Spelling Things Out

We have already pointed out the startling possibility of the voices of the dead being reheard through this device, and there is not doubt but that its capabilities are fully equal to other results just as astonishing. When it becomes possible as it doubtless will, to magnify the sound, the voices of such singers as Parepa and Titiens will not die with them, but will remain as long as the metal in which they may be embodied will last. The witness in court will find his own testimony repeated by machine confronting him on cross-examination—the testator will repeat his last will and testament into the machine so that it will be reproduced in a way that will leave no question as to his devising capacity or sanity.

—"The Talking Phonograph," *Scientific American*, December 1877

We shall be enabled *literally* to assert of Mr. Edison that, "He, being dead, yet *speaketh*, through his inventions."

—Frederick F. Garbit, M.D., Ph.D., 1878

Thomas Edison went public with his new phonograph at the New York City offices of *Scientific American*. He plunked his invention on the desk, where it greeted astonished observers and inquired after their health. After getting over their initial astonishment, witnesses reported that they were most fascinated by the sheer simplicity of the machine, "a little affair [made of] of a few pieces of metal" rather than the complicated electric and mechanical contrivance with "rubber larynx and lips" they might have imagined. The recording surface was made of tinfoil, wrapped around a cylinder that rotated by a hand crank. Indentations made on the foil formed "an exact record of the sound [waves] that produced them." These records or "remarks" could then be "translated." Although some observers thought translation could be performed painstakingly, by using a magnifying glass to discern phonetic dots and dashes, the really remarkable aspect of the device arose in "literally making it read itself."[1] Here was a kind of writing emptied of its habitual artificiality. Here was a kind of reading devoid of any literacy or any labor save the turning of a crank. Words were inscribed naturally, by the modulation of sound waves, and were recoverable as sounds without the intrusion of script, intelligence, or sight. Amid much ensuing hubbub in the press, the inventor was whisked to Washington, D.C., to give a demonstration of the phonograph at the White House for President Hayes, his wife, and members of Congress. One pamphleteer rhapsodized that a phonograph might be "placed in the interior of *Bartholdi's colossal Statue of Liberty*, now in the course of erection at the mouth of New York Harbor" to pronounce itself and liberty to foreign and domestic ears (Garbit, 15; emphasis in original). And when a huge phonograph-voiced automaton really was constructed for the World's Columbian Exposition in 1893, it was a giant Uncle Sam advertising for a shoe company—history and, proudly, American history, was being made.

With his invention, Edison made history in the banal sense of priority: he had done something that only he and maybe a secretive Frenchman named Cros could yet do. But the invention made history in an additional sense, through what Lewis Mumford called the "collective enregistration" that serves to make everyday life "more historic" (244–45). The dead could speak; the present could be stored up and never left behind. This second historicity, its imagination and desire, formed the most important cultural precondition for the phonograph. But historicizing this kind of history making is particularly difficult, for the view backward is doubly refracted by the success of mechanical reproductions and broadcast media. Mechan-

FIGURE 2. Edison and his machine in Washington, D.C. Photograph by the Matthew Brady Studio (1878).

ical reproductions clearly helped make written texts seem inadequate for capturing history. W. K. L. Dickson, for example, boasted that his and Edison's motion picture camera would supplant "dry and misleading accounts, tinged with exaggerations of the chronicler's minds" with "vitalized pictures of great national scenes" (Dickson and Dickson, 51). Broadcast media, by turn, helped make such vitalized reproductions themselves seem stale. As Roland Barthes said of the radio broadcasts during the May 1968 demonstrations in Paris, "'Hot' history, history in the course of being made, is an auditive history" so immediate that the "age-old distance between act and discourse, event and testimony, was reduced" (149). Before either movies or radio, before phonographs and televisions, making history was unquestionably done on paper by a variety of means. Graphic and legible inscriptions, like still photographs and stenographers' scripts, made history.

Making history is exactly what shorthand reporters are paid to do. Though lacking the retrospective portentousness of "vitalized" mechanical reproductions or "live" media coverage, verbatim reporting converts legislative and judicial process into historical record. It makes public memory. Even in the present age of mechanical reproduction, broadcasting, and computer networks, reporters toil away at making history. Shorthand was the subject of particular attention and acclaim during the middle of the nineteenth century, encouraged in part by the British publication of Isaac Pitman's *Stenographic Sound-Hand* in 1837. Prior to Pitman, shorthand was called *stenography* (derived from the Greek, narrow or close writing), *tachygraphy* (swift writing), or *brachygraphy* (short writing). But Pitman soon dubbed his system *phonography* (sound writing) because he claimed that his was the first shorthand based explicitly on the phonetics of English, rather than on its spelling. Andrew Jackson Graham, Pitman's staunchest American competitor, who was once his disciple, even used the term *phonograph* in 1858; according to him, the word could either be a noun, naming "the *graphic* or *written* sign of a vocal element," or a verb meaning to write using those signs (1:iv; emphasis in original). Phonetic shorthand emphasized the oral character of language at the same time that it sought to perfect a technology for linguistic representation. The inventors of shorthand alphabets saw themselves engaged in what Walter Ong might describe as "re-technologizing the word." In this manner, I will consider their alphabets as technologies—a gesture that disrupts habitual definitions of technology as mechanical or electrical, and one that will help make sense of the shorthand inventors' authorial, entrepreneurial woes.

The present chapter describes the context and history of phonetic shorthand in the nineteenth century in order to demonstrate

the experienced characteristics of textuality that preceded the phonograph. Without nineteenth-century shorthand, Edison's phonograph would not have "worked" the way it did. To Edison and listeners at the *Scientific American* offices the phonograph converted
aural experiences into authoritative, inscribed evidence. It recorded. The terms of this conversion from experience to evidence
were already in play, lurking amid the contemporary social meanings of orality and literacy. They can be discerned amid contemporary uses of phonetic scripts, the uses shorthand in government,
and the vociferous promotion of competitive shorthand alphabets.
Shorthand authors such as Pitman and Graham imagined themselves inventing technological applications within a science called
phonetics, while shorthand reporters used their technology in civil
and corporate practice. In the nineteenth century, inventors and
users shared the experienced limitations of textuality, experiences
about what should as well as what *could* be represented on the
page. Textual limits seemed pressing in several ways. The status of
authorship curiously eluded both inventors and reporters, as I explain below. Meanwhile, anxieties over the accuracy of shorthand
reports and controversies over the benefit each new system possessed hinted at other problems: the physical limitations of stenographers' own bodies as inscriptive agents; legislative ambivalence
regarding public access to debate; jural ambivalence regarding the
codification of common law; and the relative chaos of literacy in
practice, which is always dynamic as well as disruptive of rules and
a priori principles. Among the broader contexts within which the
limitations of textuality mattered were the comparison and competition of aural and visual forms, questions about human anatomy
and perception, and the ongoing definition of public against private
life, of citizens as authors and readers as consumers. Add to all of
this the sense of unease at the increase of experience in the modern
moment, demonstrated best by the ever-dilating prospect of geographical exploration, the seeming babble of immigrant voices, and
the new "human" sciences of linguistics and psychology.

The nineteenth century was far more rooted in aural experience than is easy to recover today. Aural experience was tenaciously multiple and inseparable from visual experience. In the course of the century, that multiplicity itself was the subject of repeated rearticulation. Hearing almost always came with seeing. Lectures, oratory, recitations, concerts, sermons, and revival meetings all enforced the connection of aural and visual sense in contexts that enrolled the ear within the operations of the public eye. Only the telephone (1876) and phonograph would comprise clear, popular expressions of a possibly pure and suggestively private, aural experience, presaged in medicine by the stethoscope (1816), which isolated diagnostic sound, and in specialized corporate practice by the acoustic telegraph (1844), which tapped out its Morse code. The success of such devices suggests that while the nineteenth-century observer possessed a changing technique and a shifting position, so did the auditor and the audience; like other social practices, listening was changing.[2] The changing visuality of musical performance and the noisy visibility of typing will be important to later chapters of this book. Here I am concerned with another experienced core of unity between aural and visual sense—the common but rather intricately held belief that written words were the graphic representations of speech. I stress the intricacy of this belief because contemporary experiences of phonetics were so varied: readers, who lately had begun to read quietly, "to themselves," still learned to read by so-called "pronouncing" methods. Writers, whether they worked quietly or clattered away on typewriters (1874), confronted the irregularities of sounding and spelling things out in acceptable English. And a variety of literary forms and styles evoked versions of orality. Language was insistently hermaphroditic, oral and not oral. The interest of phonetic shorthand is that its promoters set out to police certain aspects of linguistic hermaphrodism, as they went about making their livings and professionalizing their trade. In the pages below, I turn first to a description of phonetic shorthand and the competition between systems, then to the uses of shorthand in making history, and finally I place shorthand within a larger context of nineteenth-century phonetic endeavors, including the spelling-reform movement and the

nascent comparative philology, both important indexes of Anglo-
American self-identification.

~

Isaac Pitman's phonography was imported to the United States by
the abolitionist and "universologist" Stephen Pearl Andrews in 1843.
In 1852 Pitman's younger brother, Benn, followed along to publish
and promote. The next fifty years witnessed the publication of hundreds of shorthand manuals and dozens of competitive new systems, most of them "phonographic," some of them avowedly "Pitmanic." Every manual was partisan, claiming its system the most
advanced, rational, efficient, easy to learn, or quick to use. This vigorous partisanship can be daunting to impartial readers. Each edition asserts a Whiggish narrative of past error and present crowning
success. Added to this confusion, histories of shorthand were continually published by advocates who claimed their own unlikely impartiality. Even the accounts that seem the most balanced cannot be
trusted. In a late and compendious survey of the field, for instance,
the *Encyclopedia Britannica* (11th ed., 1911) alludes to recent systems as "almost entirely in the direction of script characters with
connected vowels, as contrasted with the geometric forms and dis-
joined vowels of Pitman's phonography" (1012). Thus the *Britannica* forebears mention of Gregg shorthand (1888), by 1911 one of
the strongest competitors of Pitman's phonography, particularly in
the United States. What to make of the profusion of contrary praise
and blame? A comparison to dance notation provides one helpful
parallel. (There is even a French system called *La Sténochorégra-
phie*, 1852.) Dance notation has been the subject of similar, long-
standing competition. Instructional texts are equally partisan, and
the few historical surveys equally suspect. Evaluating notational
systems, as Ann Hutchinson Guest emphasizes, depends upon acknowledging multiple variables, including personal affinity and
speed required, but more importantly, the "practical use" or context of notation (180–81). What kind and what range of movement
need to be represented, for what purposes, and for whom? Answers
to these questions involve institutional, social, and economic con-

siderations as well as artistic ones. The same sort of caveat applies to the retrospective evaluation of component technology, to the success of alternating electrical current over direct current in America, for instance, or VHS over Beta videocassettes, or IBM computing over Macintosh. The notion that the "best" technology succeeds only makes sense if "best" can involve extrinsic as well as intrinsic advantages. In the case of shorthand (as in the case of 115 volt AC, or VHS), extrinsic factors seem to have roughly followed the contours of national boundaries, since by the early twentieth century de facto national standards were in place, with Pitman favored in the United Kingdom and Gregg in the United States.

The Pitman brothers and many others invoked the origins of literacy itself in their promotion of new systems of phonetic shorthand. Benn Pitman's *Manual of Phonography* (ed. of 1860), for instance, begins with a discourse on the origins of the alphabet in the hieroglyphics of the Near East. Alphabetic characters evolved from pictographs:

No sooner had sound, instead of things and thoughts, become symbolized by pictures, than the pre-existing tendency to looser delineation became stronger and more decided, till finally, the drawing lost all resemblance to the original; and it would be difficult for us to recognize in the first letter of our alphabet, the head of Apis, the Sacred Bull. (8)

The hero of Pitman's story was the alphabet, invented by the Phoenician, whom he called "the Englishman of antiquity." The Romans then brought his alphabet to England, where it faced an impossible task: "Among the confused dialects and languages which had fused into English, many were the sounds that the Latin tongue never possessed," and for which the Roman alphabet would have to veer into barbarous and unphilosophic combinations of letters. Orthography became artifice rather than substance: English, a "strong and masculine language, slowly became invested with a garb altogether unfitted for it," dangerously cross-dressing. Phonography, Pitman promised, would return the alphabet to its substantive, heroic role, by enshrining the sounds of language in rational graphic signs. The "phonetic principle" insisted upon one sign per sound; its signs were selected from the characters of other, less perfect shorthand systems that had, nonetheless, been tested over time and sagaciously adapted by Isaac Pitman for the "great" and "wonderful" "boon" that he now "bestowed upon the Anglo-Saxon race" (11–13).

It is tempting to locate shorthand in the same broad context of the history of literacy as Benn Pitman did. The history of shorthand may indeed be the history of writing; ancient Mesopotamian representations of writing show scribes taking dictation.[3] The dubious Near Eastern "evolution" from graphical to syllabary to alphabetic writing bears a passing resemblance to the shift from spelled, stenographic shorthand to sounded, phonography, while symbol, syllable, and sound all came under new scrutiny in nineteenth-century shorthand. Moreover, frequent, determinist claims that literacy changes cognition, improves abstract reasoning, and stimulates cultural development, run parallel to claims made for shorthand as rational and scientific, encouraging mental discipline and civic progress. The alphabet is just as much a hero for anthropologist Jack Goody as it was for Benn Pitman. Goody traces the alphabet from Pre-Canaanite to Phoenician to Greek and indicates its seminal influence on economic organization and democratic government. Determinism like Goody's has been absorbed uncritically into numerous historical and literary accounts of literacy and print culture. While many authors, including Goody, have questioned the treatment of orality and literacy as stark, indivisible opposites, anthropologists, psychologists, and cultural critics persist in rating literacy according to orality.[4] Different scripts are routinely considered more or less "advanced" in their progress, with the result that the interdisciplines of literacy studies seem to offer little hope of a clear context for shorthand, unmuddied by a posteriori parameters of evaluation. As Sylvia Scribner and Michael Cole have written of the Vai, Liberian script, it seems less important and less interesting to locate shorthand systems "on some 'scale' of literacy development" than it is to understand "the social factors that operated in the past . . . to shape the contexts and nature of . . . literacy practices" (237–38), which included the use of various systems of shorthand. Rather than try to settle which is the "best" system, better to seek the variables of contemporary evaluation and the social and economic conditions that helped make those variables potent ones.

Verbatim reporting is arduous work, requiring skill and practice to achieve what shorthand authors indicate are its two essentials: speed and legibility. Casual speakers can easily utter more than 150 words per minute; shorthand reporters need to "catch and convey to paper" at comparable speeds (Munson, 122). And they need to do so in such a way that their notes can be readily understood and transcribed at a later time. Speed and legibility are twin pressures operating in equilibrium, as few writers admit. Either one gets a "tolerable degree of facility of execution with very great legibility," or one gets "tolerable legibility with the greatest facility of execution" (Graham, 1:46). The trade-off between speed and legibility gets negotiated in shorthand's economy of alphabetic characters and in matters explained as "brevity," "expedients," "combinations," or "powerful contracting principles," the ways in which characters are connected, merged, and abbreviated to save split seconds in the course of reporting. In deriding their competition, shorthand authors disparaged other systems as "arbitrary" and "mysterious," while promoting their own alphabets as "simple, practical, and complete," and their own rules as "reasonable," "practical," and "in keeping with the development of other sciences" (Graham, 1:iii–iv, passim; McKee, 11, 23). Concision, because quick, is a necessary and appealing evil. Phonography, as Isaac Pitman put it, has an inexhaustible "fund" of brevity.

Learning the reporting style of Pitmanic phonography meant learning the corresponding style first. Each character represents a sound, loosely mapped onto English consonants. Dark and lightly shaded lines and arcs are merged together into phonetic word signs. Dots, dashes, and other diacritical marks in different positions indicate vowel sounds. The word "laugh," for instance, appears as the merged characters for *l* and *f*, with the mark for short *a* beside them. Phonographic dictionaries recommend the right phonetic spelling of words. Learning phonetics sometimes means learning Pitman's own

phonetic alphabet or "phonotypy" (*Phone*, voice; *tupos*, type—
"speaking type"), a typeface designed to express all the sounds of
spoken English. Pitman even published a weekly periodical printed
mostly in phonotypy, *The Phonetic Journal*, for practice with and
appreciation of the phonetic principle. Braced with a knowledge of
"common" writing, phonographic longhand, and phonotypy, the
student could finally turn to the reporting style. Here different rules
of contraction apply. Hooks or loops represent common prefixes,
suffixes, or groups of consonants. Vowel markers are generally
omitted, the vowel sounds suggested by different lengths and positions of the usual consonants when they are not left to context
alone. And long lists of frequently used words are abbreviated to a
single letter or a single group of consonant signs. These are the so-
called "arbitraries," which some systems were more apologetic than
others about using. The sign for *p* sometimes means "up"; the sign
for *t* sometimes means "it." As the student wades deeper into reporting, contractions become more involved and patently less phonetic. Whole phrases are represented by contractions for words
composed of contractions for groups of letters.

The avowed principle at the heart of Pitmanic phonography is a
one-to-one mapping of sign to sound, in contrast to the Roman alphabet, yet in practice phonographic reporting relies upon a one-to-
many mapping of sign to sense. Homonyms and many short or similar sounding words or phrases end up having the same signs in the
hands of a skilled reporter. Every boast of "natural" can be met with
the challenge of "arbitrary." And every claim of "system" may be
qualified by "personal." Where one-to-one mapping remains pivotal,
of course, is in the conversion of shorthand reports to full transcripts,
where the unique person of the reporter and the necessary uniqueness of transcript stand in for the lacking uniqueness of phonographic signs. Though in theory any reporter can transcribe any report, practice suggested that every reporter be responsible for her or
his own transcriptions. Speed and legibility continued to be matters
of great anxiety; speed made the reporter sweat in the courtroom,
legibility made the reporter sweat later on. Instruction manuals hint

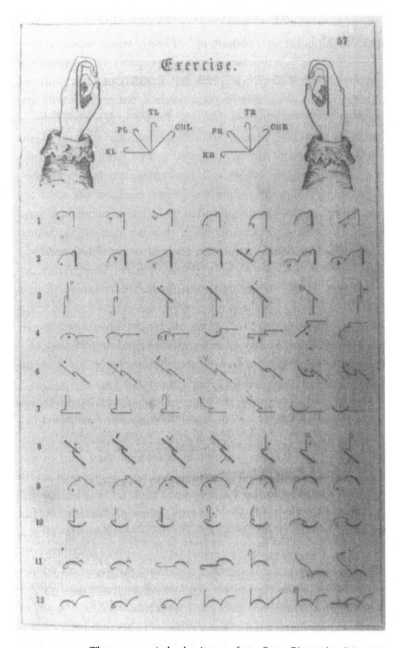

FIGURE 3. The mnemonic body. A page from Benn Pitman's 1860 manual of phonography. Pitman used the curve of a hand as a mnemonic device to help students remember his phonetic, mnemonic script. His competitor Graham used the shape of a face.

at this anxiety. The vociferousness with which rules were promulgated marks the desire to keep every practitioner of a particular system the same—interchangeable parts perhaps, within some larger
machine for turning sound into text. Paeans to shorthand as a manner of cultivating habits of memory and attention undercut the prob-
ability of achieving such a goal. Why boast that reporters have
"wonderfully improved and mechanically strengthened" their memories if following the rules means producing uniquely sensible reports (Towndrow, xi)? The conscious exertions of reporters seemed
balanced against their unconscious sensibility. Discipline vouched
for accuracy, but exertion undercut probable objectivity. Finally, reporters use different degrees of contraction, even within the same
system of phonography, and much remains to be figured out from
context while transcribing.

With steady use, a system of shorthand can be multiply personalized. Little improvements suggest themselves. Such was the origin of
so many of the new, "improved," systems of the nineteenth century.
Pitman's phonography itself passed through at least ten different editions, changing incrementally, under the supervision of a Phonetic
Council comprised of prominent reporters in Britain and the United
States. In 1851, for instance, the Phonetic Council agreed on new
consonant strokes for *w*, *y*, and *h*. The resulting ninth edition lasted
only five years; the subsequent tenth made changes to the representations of the vowel sounds. This tenth edition caused much dissen-
sion, particularly in the United States, and further stimulated the
invention of new systems by practicing phonographers. As one prescient commentator noted, Americans launched into "go-as-you-
please phonographic authorship," wherein the "exposition of an
accepted system soon gave place to the exploitation of individual innovations." For instance, the reporter Elias Longley vaunted a whole
new system on the basis of a single new consonant character
(Brown, 289). Many authors dissembled the degree of difference that
their systems bore to predecessors. Charles McKee spends pages of
his *The New Rapid: A Light-Line Connective-Vowel System of
Short-Hand Written from Phonetic Principles Without the Use of*

Word Signs or Contractions (1888) criticizing Pitmanic phonography for its disconnected vowel marks, reliance on positions to indicate diphthongs, "arbitrary" word signs and hook letters. But the *New Rapid* itself contains all of these features in some degree. Word signs, admits McKee, are not imperative, "but since there are a few words that occur very frequently, and which if written out in full would require two strokes, it is found expedient to omit one of the letters and express them by a single stroke" (64). So is this *New Rapid* very much like the old.

The vocabulary shorthand inventors used to name and promote their systems is particularly revealing. Titles and subtitles of shorthand manuals of the nineteenth and early twentieth centuries are replete with adjectives that all, in some sense, serve as antonyms for "arbitrary":

lucid	progressive	concise	eclectic
practical	improved	interlinear	simple
American	school	brief	reformed
shadeless	syllabic	advanced	"fastclear"
natural	scientific	phonetic	mathematical
English	"A.B.C."	commercial	rapid
graphic	compendious	business	"24-hour"
forensic	universal	new	complete
standard	"light-line"	parliamentary	pronouncing

Collectively, these terms, drawn from the Catalogue of the Newberry Library, illustrate the promotional rhetoric and the improving ethic of shorthand, but they also demonstrate the competitive cacophony of publications in the field. Shorthand inventors seized upon their favorite controversions of the "arbitrary" systems of others. Each title is a single gesture within the common attempt to override the artificiality of signs and, at the same time, to affirm, promote, or redeem writing as an instrument for making history and shorthand as an objective medium. This double desire for natural or "philosophic" signs and for recognition as publicly authoritative inscription was clothed by turns in a rhetoric of progressive reform, of scientific rationality, and pedagogic ease. Some adjectives, like "school," "business," or "forensic," indicate institutional purposes and aspirations. Others, like "shadeless," are descriptions of scripts, while brand names like "fastclear" and "24-hour" were probably meant to suggest tutorial ease.

In several cases new systems resembled their old models so closely that disputes erupted and litigation ensued. Like the publishers of dictionaries, the publishers of shorthand manuals vied to supply an acknowledged American standard.[5] In 1864 Andrew Jackson Graham, author of *The Hand-Book of Standard or American Phonography* (1858), sued Benn Pitman for what he reputed to be copyright infringement.[6] Graham had admittedly based his new "American" system on Isaac Pitman's ninth edition, and now he wanted to block the younger Pitman from profiting on a similarly adapted phonography. Graham swore a bill of complaint that he was the "author, originator, and composer" of new "principles, rules, combinations, devices, contractions, and word-signs" he had presented in a "new, regular, and scientific order and method." (Graham likely believed that the Englishman Benn Pitman was unable to secure effective American copyrights for his publications.) Pitman testified at length and seems to have demonstrated convincingly that Graham's phonography was no unique "standard." As many others also found out, American copyright law does not protect shorthand systems any more than it protects bookkeeping systems or computer-programming languages. Only the individual printed expressions that explain a system are protected. For all his fuss, Graham's gains were purely rhetorical and significantly nationalistic, and a decision in the federal copyright case was never recorded. In Britain, where Pitman forces were stronger and copyright law differed, Isaac Pitman successfully blocked the sale of both Graham's *Hand-Book* and another American upstart, James Munson's *Complete Phonographer.*[7]

Similar controversies dogged John R. Gregg throughout his career. First he was sued in England over the very system of "light-line script phonography" that he had been instrumental in developing.[8] Many of the heavyweights of late-Victorian shorthand provided affidavits against him, including George Carl Mares, who swore that Gregg's light-line phonography and the plaintiff's script phonography "are exactly the same," with the exception that Gregg omits "the thickening of signs to indicate the letter S and the device of indicating the letters L and R by the position of other signs." Gregg's side of the case was left to argue that these differences were not trivial, and that Gregg's "absence of shading" and "absence of 'position writing'" were original and significant enough to constitute a whole new system. After obtaining a preliminary injunction, the Script Phonography Company did not pursue its case, which was dismissed out of court with costs to the defendant. When he immigrated to the United States in 1893, Gregg solicited legal advice and followed case law closely, trying to protect himself against pirates and competitors. Over the years he weighed copyright infringement, patent infringement, unfair competition, and trademark violation as the possible grounds for prosecution. Though Gregg realized that patent law offered no protection, it is not impossible to find other shorthand manuals with the words "Patent applied for" printed on the reverse of the title page where a copyright notification might regularly appear.[9] Trademark law proved the most applicable. As his system gained in popularity John Gregg contented himself that his own name was "the greatest protection of all against plagiarism of the system"; no one else could publish a system called "Gregg" or "Gregg Improved." After he was famous, Gregg received letters of apology from Mares and others.

Why such tumult and controversy? What were the stakes that made a few shaded letters or a single new character seem so important? What were the social and economic factors behind such a chaotic indemnification of "arbitrary" signs? The authors of shorthand systems usually published their own works: Benn Pitman had his Phonographic Institute in Cincinnati; Graham had his Phonographic Depot in New York; and later Gregg had his Gregg Publishing Company in Chicago and then New York. This may initially have had to do with the expensive specialization of type fonts and engravings as well as the typesetting and proofreading skills required to print shorthand, but the corporate organization clearly arose from a particular menu of possible economic relations as well. Shorthand publishers made money in five ways. They sold their manuals directly to students. They sold their manuals indirectly to students, through the shorthand schools they established. They collected tuition for study at the schools. They collected tuition for correspondence courses. And finally, they provided skilled reporters for hire, their schools acting as clearing houses and employment bureaus for graduated students.[10] It was a text-based economy without authorship, in which copyright didn't matter, couldn't matter, and allegiances were everything. With intellectual property moot, mental exchange loomed large. Agents were granted exclusive territorial rights to sell books in exchange for signed agreements requiring them to establish schools and promote the system. The Phonographic Institute in London offered the stereotype plates of its books and pamphlets to Americans for relatively modest sums, if only responsible parties would accept exclusive territorial rights in exchange for promoting Pitman's phonography. A board-bound manual cost less than a half dollar through the 1840s; by the late 1880s only the most sumptuously bound handbook cost two dollars.[11] With prices like these, entrepreneurial eyes were not only on the book trade but also on the educable labor market. The aim of course was to create an ever-growing base of customers wanting to learn just that system.

Shorthand publishers competed for market share the way periodical presses competed for circulation and advertising copy. Like the publishers of small-town newspapers, shorthand authors probably generated copy, handled elements of the printing process, and attended to promotion as well as other business matters. (And they took reporting gigs to make ends meet.) But the analogy is imperfect. Shorthand publishers resembled the purveyors of new, component technology as much as they resembled newsmen. They needed to amass market share, but the rewards for doing so were more than purely arithmetic. Under the market's own pressure for standards, an increasingly larger share could finally tip the marketplace toward one system or another, yielding an exponential reward. In this way the marketplace has tipped toward a single, standard railway gauge, a standard electrical wall outlet, alternating-current rather than direct, disk-shaped phonograph records rather than cylindrical ones, VHS rather than Beta, and, gradually, OS computing with a "graphical interface." As different as these examples may be, standards make components fit, no matter who makes the components. One of the lessons of shorthand is that literacy skills can act as components within the broader context of literacy practices. "Literacy," like "The Market," tends toward standards.[12] Shorthand alphabets mimic network technologies; shorthand itself was a nineteenth-century "dynamical system . . . in which local *positive feedback* mechanisms predominate over negative feedback mechanisms, and which can therefore be characterized as 'self-reinforcing'" (David, "Heroes," 133, emphasis in original). Success breeds success.

Two classic examples of technologically implicated standardization in Western literacy practices offer different and valuable points of comparison: the standardization of orthography with the early-modern use of moveable type, and the tenacious hold of the QWERTY keyboard design on typewriters and computers. About the former there is no controversy. Authors as different as Henri-Jean Martin, H. L. Mencken, and Benn Pitman have long recognized the diffusion of printing acted as an orthographic buffer, gradually dampening volatility and smoothing out variations. Particularly when contrasted with pre-Gutenberg spelling, printed English seems to have frozen in its tracks about the time of Shakespeare and the King James Version of the Christian Bible. This standardization arose from technology only in the sense that the printing press allowed a greater production of text, which in turn allowed a greater diffusion, implying an increase in literacy rates that involved a newly standardized literacy in which "spelling counts."[13] The author-publishers of shorthand manuals sought to standardize writing in a similar fashion, through printing, diffusion, and education. Each inventor wanted the rules of his system to become standard, but with so many different systems in competition it may have been difficult to convince individual students that any rules were hard and fast, that shorthand had (or even needed) mature normative characteristics.

The fortunes of the QWERTY keyboard may be even more pertinent. Paul David explains the tenacity of the modern QWERTY keyboard as a matter of economically determined error. The QWERTY arrangement of the original Remington typewriter has remained virtually universal since the 1890s, even though more efficient arrangements have been developed. The market has tipped to the wrong standard. Remington's arrangement was based on that of the inventor Christopher Latham Sholes, who wanted to keep the typebars from clashing when the operator typed quickly. Better-engineered machines and then electrics and then computers vitiated the need, yet QWERTY stuck (in parts of Europe, AZERTY). (Reportedly the Remington Company also liked its product name, "type-writer," to appear acrostically in the top row.) David's work has been attacked by S. J. Liebowitz and Stephen E. Margolis. Liebowitz and Margolis challenge David's account of the facts, particularly his identification of a proven-better keyboard, but they also attack his model of the market. Their market is less prone to error. They quickly sketch the early history of market competition and suggest that QWERTY succeeded because it *was* the best, implying that it may have been easier to learn than its rivals, since ease of tuition could be just as important as ease of use. In the end there is less difference between David and Liebowitz and Margolis than the latter make out. Both parties admit the pressure of standards, even if Liebowitz and Margolis hold that standards are more sensible. All ascribe the conservatism of standards to economics: retooling and retraining make design changes prohibitive.

My sympathies are with Liebowitz and Margolis's facts and David's view of the market, which are not necessarily contradictory. The former's description of market competition is forcibly reminiscent of shorthand:

In the 1880s and 1890s typewriters were generally sold to offices not already staffed with typists or into markets in which typists were not readily available. Since the sale of a new machine usually meant training a new typist, a manufacturer that chose to compete using an alternative keyboard had an opportunity. As late as 1923, typewriter manufacturers operated placement services for typists and were an important source of operators. . . . Manufacturers internalized training costs in such an environment, so a keyboard that allowed more rapid training might have been particularly attractive. (19)

So too did shorthand publishers need to cultivate the labor market in order to disseminate their systems. But the publishers never internalized cost. Though there may have been on-the-job training for typists or free typing classes and job placement for typists, little was ever free in shorthand. Rapid training had rhetorical appeal more than it had cost effectiveness. Even the quickest system, if one emerged from the shorthand contests of the later century, like the quickest keyboard, might succeed or fail for any number of reasons, particularly if "quickest" was a matter of inches, as it appears to have been in several documented cases.[14] Among the pertinent variables of success were geography, institutional vigor, marketing skills, and access to literate, educable labor, and a variety of target markets. It is impossible to know in retrospect whether or to what degree Gregg shorthand eventually succeeded in the United States because it was quicker or easier, more rational or practical, or because John Gregg most successfully oriented himself near centers of power, within trademark law, and amidst a vigorous and flexible contractual network of book suppliers, employment providers, and educational agents.

Even if the market settles on the best standard, it may settle for the wrong reasons. Or "best" may be entirely a construction of hindsight in ways that are difficult to recover. Certainly one portion of any retrospectively "best" keyboard or shorthand system is an element of design commitment that validates economics, a complex emotion really, which the design engineer Donald Norman calls the "psychology of everyday things" (145–51), and which is as much rooted in the satisfactions of specific, literate skill acquisition as it is in unacknowledged economic expedience. I am suggesting that the past standardization of English orthography, keyboard design, and the gradual emergence of national preferences in shorthand systems shared certain individual and collective emotional and psychical rewards, in addition to the obvious economic benefits available to typewriter manufacturers and shorthand publishers. These rewards follow the collective accedence to any standard, the individual acquisition of any skill. But they also accrue to any able participation in literacy practices, which interpenetrate the broader power relations of a literate society. The prohibitive expense of retooling and retraining that conserves the QWERTY keyboard finds a compliment in the conservative qualities of literacy practice. Admitted normative characteristics make spelling, for instance, a matter of right and wrong. Misspelling is transgressive and signals the speller's marginal status, either preeducated, uneducated, or sloppy. So knowing and complying with a specific shorthand system, amid its rigorous structure of approved rules and its vociferous promotion of "best" systems, must have lent its users a sense of rightness, of authority, of being in step, which comprises so much of the ideology of literacy. Many disparate elements of different literacy practices are normative, notwithstanding the appreciable liberation and empowerment rightly ascribed to literacy as a whole.

~

While shorthand inventors and reporters sought to promote their own rules as the best or correct means of representing speech, they equally sought to promote the use of shorthand reporting itself as a professional and necessary means of controlling public memory. They promoted their technology as an objective medium; the term they used most frequently to vaunt its objectivity was "verbatim." Shorthand had arrived in the nineteenth century as an already tested crucible for private and public discourse. Many early English systems were developed by individuals seeking to rationalize, speed, or even encrypt the private, occasional matter of diaries or letters, and who then might commend their systems to others in print. Capturing sermons was another frequent use, an appropriation of public speech into private hands. The tangle of cross-purposes survived well into the 1860s and 1870s. Pitman's phonography had two different styles, a "corresponding" style and a "reporting" one. Pitman aimed his corresponding style at "the merchant, the lawyer, the editor, the author, the divine, and the student," that is, at the private use of affluent male auditors. By contrast, he intended his second style for the use of verbatim reporters, a class that included newspaper and police reporters as well as court and parliamentary reporters. Unlike the correspondent, the reporter was a technician, less affluent and, as the 1870s and 1880s progressed, occasionally female. The correspondent used shorthand for private edification; the verbatim reporter used it for the public record.

According to Ian Hunter and other scholars, the very concept of *verbatim memory* is limited to literate societies. The idea of text determines verbatim recall at the same time that it offers the only possible mnemonic check.[15] One can only know if one has something "by heart" if one also has it written in hand. By a similar token, the preparation of lengthy, word-for-word transcripts of oral proceedings only became a practical possibility with second-generation or a posteriori literacy; that is, with writing that abbreviates or otherwise speeds up what nineteenth-century authors of shorthand manuals were so fond of calling "common" writing. The terms *first-* , *second-* , and *third-generation literacy* come from computer programming, where written programming languages build upon each other the way shorthand builds upon "common" writing.[16] According to Sir Edward Maunde Thompson, the Greeks may have had a second-generation language as early as the fourth century B.C.E. A marble slab discovered on the Athenian Acropolis appears to be the first shorthand manual, its inscription describing a rudimentary system for abbreviating Greek orthography.[17] Pitman's phonographic reporting style is a third-generation language in many respects, based on his corresponding style, which is based on the sound of English. It does seem to have offered the first widespread, reliable system for verbatim reporting; its several offshoots became the systems of choice for parliamentary and court reporting in the United States and Britain.

In Washington, D.C., political reporting was handled by newspapers and other for-profit publications during the eighteenth and most of the nineteenth century.[18] The *Congressional Globe* was not verbatim and diurnal until 1848 for the Senate and 1849 for the House, when Congress embraced verbatim shorthand reports. The varying incarnations of the *Globe*, which was succeeded by the government's *Congressional Record* in 1873, indicate that reporting legislative action was far from easy to settle. Preceded by such publications as *The Debates and Proceedings in the Congress* and *Register of Debates and Proceedings*, the *Globe* carried a descriptive subtitle, "Containing sketches of the debates and proceedings, . . . " until the 32nd Congress (1853–1854), when it became more bluntly "Containing the debates and proceedings." The essentialist nature of published text seems questioned, title to title, and it was not until the arrival of verbatim reporting that text safely *equaled*, rather than "registered" or "sketched," the aural experiences of debates and proceedings. As it turned out, the debates of individual congressional committees were not reported in full until 1866, and reporters were not made officers of the House until 1873, notwithstanding the fact that reporters had been allowed access to the House from its very beginning, because of an admitted (if occasionally lamented) equation between public access, print, and democracy in the constitution of an American civil society. By 1879 congressional reporting cost taxpayers $50,000 per year, half for five reporters in the House, and half for a contractor who subcontracted reporting in the Senate.[19] One reporter later described the peculiarities of capturing congressional debates. As soon as a member began speaking, the reporter dashed toward the speaker, writing as he went:

This "sprinting" feature of shorthand work is a little novel to me. Many an innocent little remark which smiles from the pages of the "Record" has been literally run down on paper by the combined energy of fingers and legs.[20]

"Running hand" was a feature of nineteenth-century penmanship as well as shorthand, and this "literal" running down of legislation by hands and legs offers a counterpoint to the usual corporal metaphors of government, legislative bodies, and heads of state.

Chambers included a specific place for reporters when they were not running around—a table at which also sat a reporter from the Associated Press wire service. Newspaper reporters in the gallery kept running too, not toward the members of Congress but away from them, in the direction of the Washington telegraph offices. Shorthand reporting thus formed a web of inscriptive action, making the work of Congress public record by doubly making it public and making it record. A similar web had existed before 1848 and 1849, but the promotion and spread of verbatim shorthand reporting made the weave closer and changed the way its patterns appeared.

Prior to Pitmanic reporting the only elements of American legislative proceedings to receive consistently detailed treatment were the texts of bills and the prepared speeches delivered by representatives and senators who handed a copy over to the congressional clerks for inclusion in the *Globe*, or who published copies themselves for franking to constituents. Speeches that were otherwise "taken down by hand" were often shown to their speakers for correction before publication, where they were cut and spliced into the newspapers of the region. American government thus passed into history smacking of oratory, rather than debate, of issues and positions, rather than exchanges, and of arrangements, rather than events. That made democratic government different. Certain details of congressional hearings, for instance, which are today such a fundamental part of civic life, could seldom reach the public eye and inflame the public imagination. While television, not shorthand, is responsible for the immediacy of the McCarthy and Anita Hill / Clarence Thomas hearings, immediacy is not the whole story. What does not or cannot exist as record cannot be made immediate. Before the adoption of verbatim reporting there was less opportunity for hearings and debates to matter, literally, to comprise the material records of governance. They were narrated more properly than they were quoted. History was generally described, not made, on paper. Readers of the *Globe* found the legislative process premeditated more than spontaneous, giving center stage to the authors of texts, written bills, and prepared speeches, punctuated by dramatic

vignettes. The adoption of verbatim reporting signaled a shift taking place in the relations of authorship and citizenship: "Congress"
was now more properly the author of the congressional record than
its individual members were, as more and more shorthand reporters
were engaged in making history. If the shift gradually made authorship and authority broader, more abstract, and more disembodied,
then it also made the mechanics of government a little more material, a little more embodied in a textual sense.[21]

Too much textuality, like too much immediacy, can be a bad
thing, or at least a complicated one. As the Senate reporter D. F.
Murphy explained to the first International Shorthand Congress,
held at the Geological Museum in London (1887), "The debates are
published *verbatim* in one sense, but of course not so as to present
glaring grammatical inaccuracies or slips of the tongue; they are *not*
condensed, or edited for style" (51; emphasis in original). "Verbatim" had its own shades of meaning in parliamentary reporting, and
Murphy seems to have sensed no contradiction between editing for
mistakes and editing for style. There were powerful reasons for
wanting things both ways. Reporters and reporting lost a lot of
their cachet if transcripts were condensed, yet editorial emendations
also comprised evidence of the reporter's intellectual abilities, discursive talents, and continuing good will. Politicians notably shared
the same ambivalence toward immediacy and inscription. The inventor Thomas Edison reportedly came to Washington in 1869 to
promote his first patented device, a vote-recording machine that
would immediately and electrochemically register the votes cast by
individual members of the House and Senate. The machine was
considered but never adopted, and Edison and his biographers subsequently quoted a committee chairman, "Young man, if there is
any invention on earth that we don't want down here, it is this"
(Dyer and Martin, 102).[22] Too much disclosure often proves incon-
venient. If D. F. Murphy denied that congressional transcripts were
"condensed," then a term with even more problematic force was
"digested." Prior to verbatim reporting, debates were published as
digests, if at all. The new reporters did not digest; instead, proceed-

ings passed through them verbatim, from aural experience to visible inscription.

Like the proliferation of opposites for "arbitrary" in the promotions of shorthand inventors, the semantic range of alternatives for "verbatim" among "edited," "condensed," and "digested" suggests an anxiety over the status of shorthand reports as representations. Reporters wanted to be accepted as objective, even mechanical, agents of representation. Yet neither they nor the government they reported found such authority all that easy to assume. The matter had come up with force before, most notably during the very first federal Congress in 1789, and then during the impeachment trial of President Andrew Johnson in 1868. In the first federal Congress, Revolutionary War veteran Thomas Lloyd and other newspaper reporters had competed to supply the periodical press and their own printing endeavors with copy. Using his own modified shorthand system, Lloyd reported, transcribed, and published the first *Congressional Register* of thirty-five weekly issues, which he advertised as "Containing an impartial account of the most interesting speeches and motions; and accurate Copies of remarkable papers laid before and offered to the House" (qtd. in Tinling, 527). But impartiality and accuracy both proved nettlesome, as did Lloyd's judgment of the "most interesting" oral proceedings. Many critics condemned Lloyd's federalist bias, and members of the House, including James Madison, made pointed remarks about the powers of imagination he exerted in rendering his reports. The House debated whether to ban reporting altogether, but took no action.[23] Then in 1868 the Senate returned to the matter, taking up the truth claims of Pitmanic phonography in its inquiry into statements attributed to President Andrew Johnson. Dubious distinctions were drawn between reporting from sense and reporting from sound, and several reporters were asked to testify. Reporter James Sheridan was questioned closely about phonography:

Q. Which is, as I understand, reporting by sound and not by sense?
A. We report the sense by the sound.
Q. I understand you report by sound wholly?

A. Signs.

Q. And not by memory or attention to sense?

A. No good reporter can report unless he always pays attention and understands the sense of what he is reporting.

Q. That is the very point I wish to arrive at, whether you are attending to the sound and setting it down in your notation, or whether you are attending to the sense and setting it down from your memory or attention to the sense?

A. Both [. . .]

Q. Your characters are arbitrary, are they not? That is, they are peculiar to your art?

A. Yes, sir.

Q. They are not letters?

A. No, sir.

Q. Nor words?

A. We have word signs.

Q. But generally sound signs?

A. We have signs for sounds just as the letters of the alphabet represent sounds.[24]

Sheridan's bungling interlocutor somehow posits three mutually exclusive means of knowing—sound, sign, and sense—only the last of which he wishes to validate as the true work of memory. But with his vocational understanding of signification as a process, Sheridan can only overlap all three. After the confusion of this exchange, another reporter, James O. Clephane, vouched for the accuracy of shorthand reports in terms that seem similarly problematic. Clephane was certain that phonetic shorthand resulted in an "accurate, a literal report," with the qualification that "a word or two here and there" get changed in order to "make the meaning more intelligible, or to make the sentence a little more round." Like objectivity in other professions, objectivity in reporting relied upon a kind of neutral, or impartial, accuracy apparently muddied by the agency, by the consciousness, sapience, and intentions of the reporter. Reporters, who relied upon the patronage of legislators, wanted to make their motives seem pure and their patrons look good. Even in the courtroom, court reporters were instructed to use "good judgement" in reporting the comments of trial lawyers and judges, because even the best speakers do not "speak extemporaneously in such a manner that they would be willing to see a verbatim report of their words in print." This contrasted with the testimony of witnesses, whose "exact language, whether grammatical or ungrammatical, should be preserved" (Munson, 125–27).[25] Yet the conscious exertion of skill might abnegate the unconscious conduct of accuracy. Heated exchanges took place over the admissibility of shorthand reports as exact evidence "once and for all." They were accepted largely on the basis of past use, and the hearings continued to other witnesses after lamenting the absence of manuscript sources in the place of reports. Reports were going to have to be textual evidence, at least until some better text, some ultratext, turned up.[26] It was perhaps twenty-two-year-old Edison's misfortune that he tried to promote his vote recorder on the heels of this debate.

In jurisprudence as in governance, the increasing use of shorthand interrogated the textual representation of aural experience. According to a survey done by the New York State Law Stenographers' Association, New York was the first state to introduce shorthand reporting into the procedural rules of its legal practice. In 1860 the state declared that judges could require shorthand reports, the costs to be shared by the litigants. Litigants had long been able to hire their own reporters if they desired, but now the state could require it. Other states and municipalities followed New York's lead during the following decades, and shorthand reporting gradually became an integral part of legal proceedings. If deeply ambivalent about the codification of common law, the American legal system was already firmly committed to textuality. It had a statutory base, followed procedural codes, and was articulated according to precedent, which accreted in the annals of case law via the reports of judicial decisions. All these attributes may again be unique to literate societies;[27] without the notion of verbatim memory, "preliterate" societies exist in a dynamic soup of tradition, as present judgments slip into hazy precedents and are released into the habits of cultural identity. By contrast, American jurisprudence cultivated public memory by producing and perpetuating text. Court decisions formed its preeminent genre, but lawyers' briefs were also written and often printed, while documentary evidence, affidavits, and the "writs" of legal process, instruments like subpoenas and injunctions, served pointed textual needs. What remained in the dynamic soup of orality were the proceedings of trials. In particular, clerks digested the examination and cross-examination of witnesses; fragments of testimony could be hazily recapitulated in decisions, arguments, or briefs, but testimony had little status of its own as a matter of record, until shorthand reporters took it all down.[28] Imagine Perry Mason or Court TV without "Let the record show, . . . " or without challenges to witnesses who testified to one thing and later changed their stories in some small way. Verbatim reporting offered a way to recapture more of the visuality of witnessing, because oral accounts could fully enter the disciplinary operations of the court as matters of record. It affirmed changes taking place within civil notions of proof and surety. Testimony could be scrutinized in new ways, saved, compared, and faithfully quoted as never before. Similarly, a judge's performance might be evaluated with new authority as incidental comments and directions from the bench entered the record verbatim. The appeals process could become differently exigetical. And another part of Barthes's "hot" history could newly become the written object of cool reason.

The rules of evidence hadn't changed, but now there was so much more evidence of different things. It was an "explosion of fact," as anthropologist Clifford Geertz observes in a comparable instance, with a corresponding "fear of fact" evident in the lengthening rules of procedure, concurrent debates over the codification of common law, and a confusion over the rights and status of authorship with regard to court reports (171).[29]

Court reporting provoked uneasy questions about who authors were. Early law reports had carried the name of their reporter, either a stenographer or a judge (e.g., *Wheaton's Reports*), with the result that "Names like Wheaton, Dallas, and Howard, are all better known than some of the Nineteenth Century justices of the Supreme Court"; yet by midcentury such reports were cited by the name of the state or court (e.g., *North Carolina Reports*), suggest-

ing a growing awareness of civil authorship and corresponding amnesia about the reporter as vehicle.[30] These same questions had been
hammered out by the Supreme Court in the case of *Wheaton v.*
Powers (1834) over the matter of Henry Wheaton's copyright on
Supreme Court reports he published between 1816 and 1827. In
holding that Wheaton's copyright protected only the reporter's editorial apparatus, not the reports themselves, the court "exempt[ed]
the law itself from the rules of authorship it sets in force," ostensibly rendering the court's own report as an ideal and essential text,
"one that operates powerfully in the absense of an author" (McGill,
42).[31] Courts *made* case law; reporters didn't author it. Authorial
absence may have disappointed reporters, but it did not stop them
from trying to profit from the publication of celebrated cases that
formed a popular genre among nineteenth-century readers.[32] Nor
did it, in granting their text its supreme, unauthored status, mitigate
the bibliographic habits of publication and the marketplace. Benn
Pitman reported the two most famous cases of the postbellum moment, the trial of the Lincoln conspirators in 1865, and the Ku Klux
trials in 1871, both of which tested the resolve of Reconstruction.
The transcripts of both cases survive on general library shelves today, and their bibliographic records indicate unstable authorial parameters. The titles of the two books are *The Assassination of Pres-*
ident Lincoln and the Trial of the Conspirators: The Courtroom
Testimony as Originally Compiled by Benn Pitman, and *Proceed-*
ings in the Ku Klux Trials at Columbia, S.C., in the United States
Circuit Court, November Term, 1871. The former, published originally by a commercial press in New York, has been given a more
bookish, literary title, while the latter is a facsimile edition retaining
the jural nomination of the government original.[33] The first author
of the Lincoln book is David E. Herold, one of the ill-fated defendants. The other specified authors consist of Mary Surratt and Dr.
Samuel Mudd (two other defendants), reporter Benn Pitman, and
the Military Commission of the U.S. Army, which presided. The primary author of the Ku Klux volume is the presiding United States
Circuit Court, though Benn Pitman and his associate Louis Freeland

Post are also listed as authors. Defendants, reporters, and courts all authored proceedings in this era of court reporting. Authorship is scattered, a matter of simultaneous orality, inscription, and author-ity—who speaks, who writes, and who presides—because all three were necessary, as one shorthand author put it, "to sustain to the *light* and *sight*" what were otherwise fleeting matters of "the *air* and *ear*" (Graham, 1:9; emphasis in original).[34]

~

Despite the elusiveness of their authorial status and the indistinct-ness of their (disdained) "arbitrariness" and (desired) "verbatim" reporting, shorthand inventors and reporters proved successful in promoting themselves and their profession. Association with law, government, and business granted shorthand a certain facticity, and it became enrolled within the rationalist and technocratic ideals of the mid- and late nineteenth century. After the first meeting in Lon-don, international shorthand congresses were held in conjunction with the international expositions in Paris (1889, 1900), Munich (1890), and Chicago (1893). Shorthand jumped on the bandwagon of industrial progress, technological achievement, national excel-lence, and the purposeful self-possession of Western colonizing pow-ers. For individuals, education in shorthand offered self-reliance, self-discipline, clean hands, a white collar, and an objective vehicle for the representation of plain speech and a knowable world. At the World's Columbian Exposition in Chicago there was even a Stenog-raphers' Day in honor of the cohort. But shorthand was not the only or the most widespread means in which phonetic inscriptions were used to make history or delimit its subjects. The acceptance of short-hand as a vehicle for public memory was part of a larger context within which nineteenth-century writers and readers had to tackle questions about the correct—or even possible—textual representa-tion of linguistic sounds. Spelling formed the largest part of that context, as individual students wrestled with the intricacies of En-glish-language orthography and as authors, reformers, linguists, and travelers all differently wrestled with the range and implications of linguistic difference.

Though relatively few Americans had any direct experience with shorthand, literates all had some contact with the underlying matter of representing orality. Children labored noisily over spelling books.[35] Readers consumed published oral forms—lectures, sermons, and trial reports—while the more fictive genres of American literature frequently appropriated or invoked the orality of yarns and tall tales. Literary authorship required a good ear, most notably as it was practiced by Mark Twain or John Gregory Dunne, or by regionalists who relied upon the exaggerated verisimilitude of printed dialects. In such contexts orality signaled identity and otherness, both the raw and regionalist otherness of unlettered or unrefined speakers, but also the otherness that separated individuals, nations, races, classes, and cultures severally from one another and the "sociological tension" that accordingly held them together (Simpson, 145). In his "Explanatory" at the beginning of *The Adventures of Huckleberry Finn*, Twain boasted about the "shadings" of dialect he had "painstakingly" represented, only half joking when he warned his readers not to suppose that his characters "were trying to talk alike and not succeeding." His printed dialects, like Mr. Dooley's, were a matter of spelling and punctuation, and his "Explanatory" is a defense against the possible charge that they were arbitrary, not accurate. That is, Twain's claim to have discerned differences in "shading" between the sounds of regional dialects is equally a claim about his proficiency in spelling them. Orthography was one ground upon which literate English speakers negotiated their own identity and the identity of others while at the same time experiencing writing as artificial, glimpsing everywhere the potential failure of textual representation to recuperate aural experience.

Though space will not allow anything like a complete discussion of nineteenth-century linguistics and its relation to phonetics, I do want to allude to two important points of connection, the spelling-reform movement and the emerging discipline of comparative philology.[36] In Britain Isaac Pitman became an unswerving advocate of spelling reform, plotting and propounding the improvement of the race by means of the gradual transformation of spelling into phonetics.[37] In the United States, meanwhile, spelling reform possessed a lingering nationalist flavor. Ben Franklin and Noah Webster had both wanted to rationalize spelling, which they took to be a particularly American gesture. Spelling-reform bills were put before Congress in 1844, 1878, and again in 1888.[38] The organized Simplified Spelling Movement tapped all of these sources and gained momentum alongside other reform movements of the era. In 1876 the International Convention for the Amendment of English Orthography met in Philadelphia, where the Centennial Exposition was held the same year. The resulting Spelling Reform Association gave way to the Simplified Spelling Board, which spent some $283,000 of Andrew Carnegie's money promoting the cause between 1906 and 1919. Despite the brief flirtation by Teddy Roosevelt's administration, the movement went nowhere. Even its own advocates frequently printed their tracts and appeals in conventional spelling, damned on their face by an obvious case of what social scientists call the *reflexive problem*. So did Isaac Pitman long and curiously suffer the *ph* in "phonography" without attempting plain "fonografy."[39] Having achieved a standard form, spelling is at least as hard to budge as QWERTY on a desktop.

At the same time a matter of individual pedagogic encounters and failed social reform, the potential of spelling as form of phonetic representation was also explored in the new and various science of linguistics. In Britain and the United States nineteenth-century linguistics combined multiple strands, mixing the British antiquarian tradition, a lingering scholasticism, and imported German romantic philology under the dual pressures of increasing disciplinary definition and continuing global exploration, which relentlessly turned up new questions for study. In America these strands were further entwined with the study of indigenous languages and the exploration of differences emerging between the American and British forms of English. Like shorthand inventors, philologists wrestled with questions of aurality and visuality, of experience and evidence, that were not unrelated to the broad underpinnings of Anglo-American self-identification. Along with the epistemics of imperialism, British and American self-identification came loaded with assumptions about language and about inscription. Philologists sought linguistic history by looking closely at English, while comparativists sought the same history abroad. Within philology there were some who construed a crucial distinction between phonology and etymology, between hearing and seeing. Alexander J. Ellis made the point with assurance. British spelling reformer, philologist, pioneering musicologist, and later translator of Helmholtz, Ellis published the first part of his *On Early English Pronunciation with Especial Reference to Shakspere and Chaucer* in 1869. Despite his literary antiquarianism, Ellis rejects etymology and emphasizes the romantic organicism of spoken language. "A spoken sound once written ceases to grow," he puts it; "Each written word is, as it were, but an instantaneous photograph of a living thing, fixing a momentary phase, while the organism proceeds to grow and change" (17). Written words, either by Chaucer or Shakespeare, were of philological value largely as the material means by which scholars could ascend to the oral forms and aural experiences of the past, recapturing the animation, the vitality of utterance wherein true Anglo-Saxon identity resided. Ellis's photographic metaphor is a break from the unrelenting geologic metaphors of so many contemporary philologists. Under the sway of uniformitarian geology, philologists tried to see the past buried in the present, while Ellis, a close associate of Isaac Pitman, seemed sure he could *hear* the past if he looked hard enough.

The second strategy that the nineteenth-century possessed for the study of the linguistic past involved the common and insidious identification of remote peoples with equally remote histories. Glimpsed everywhere, primitivism uncritically remained an argument for cultural evolutionism. Philology was enrolled within the practice of ethnography to lend credence to different narratives of population. Families of languages indicated descended or ascended families of man. Explorers were instructed to collect languages the way they collected new plant species and meteorological data. The "vocabularies" they compiled stood intermediate between collected specimens and collected data. Collected words comprised individual specimens, samples of foreign tongues, but unlike botanical or zoological specimens, words could themselves be included in the text of an explorer's report or a missionary's account. Collected words were a perfect resolution of fact and ethnographic artifact, and the explorer's "vocabulary" consequently formed a kind of museum, a repository where collected materials could flatter the scientific interests of colonialist observers at home and in the field. Like shorthand instruction books, vocabularies served as mute evidence of themselves, attesting to the preeminence of inscribed evidence in the repeated shuttle between aural experience and observable fact. Global exploration forced upon the West the recognition that hundreds of "new" languages existed, each a seeming welter of confusing new sounds and connected to newly glimpsed sites and speakers. British and American explorers and functionaries could substitute English for native languages as the lingua franca of colonialism, but in the "vocabulary" they possessed a more subtle tool for the objectification of indigenous tongues and dialects. Inherent in the publication of these vocabularies were the assumptions that, first, non-Western words could be rendered into phonetic English, and, second, that English formed a template of essential meanings into which all non-English words could be fit in a one-to-one correspondence. Lieutenant Charles Wilkes, commander of the United States Exploring Expedition (1838–1842) was simply handed printed lists of English words by the War Department, so that copies could be "filled up as circumstances permit" with lots of "Indian" languages (1:xxxi). Thus, in a genre notoriously lacking in information concerning communication between Western explorers and non-Western people, which must have been difficult at best, collected vocabularies offer a small part of the picture exploration accounts leave so obscure. A few word collectors even admitted the difficulties of the project, acknowledging that many alien phrases for "What do you mean?" had likely been dutifully written down as the "denomination of some animal or thing" (Stokes, 2:23). Here was the danger that Western travelers would be caught in an antilanguage, the way the Cyclops gets caught calling Odysseus "Nobody."

If the rhetorical problem of the explorer's account was to recount what he had seen and heard in a believable manner, then the additional burden of *representing* what he had heard took a decidedly subsidiary role. Explorers were crippled by their own lack of philological training and by the absence, even in philology, of a standard phonetic alphabet for the representation of pronunciation. Most simply assumed that English would do. Charles Wilkes, for instance, writes that the language of Samoa is "soft and smooth"; "The letters that the missionaries have found necessary to adopt in order to write it are only fourteen in number: *A E F G I L M N O P S T U V*" (2:123). Such a system proved inadequate, though, when the missionaries tried to pronounce what they had written: they were "liable to make many mistakes which appear absurd to the natives"! Other explorers and other missionaries proved far more resourceful, but there was no standard. Alexander Ellis proposed a system of his own, "palaeotype," but called it a "makeshift scheme," no matter its utility to "indicate the pronunciation of any language with great minuteness and typographical convenience" (1). To palaeotype he immediately added "glossotype," a system of two diacritical marks to meet the requirements, he said, "of writers of our provincial dialects" (13). His addition seems to confirm H. L. Mencken's wry observation that even after a late-nineteenth-century standard was adopted in the International Phonetic Alphabet (IPA), new languages always seem to expose defects in the system.[40] Wilhelm von Humboldt, upon whom many later authors relied, had spent his final years obsessively revising his linguistic and phonetic systems, each revision provoked by his acquisition of a new language.[41] His phonetic representations remained beset by alien sounds that continued to freshen the problem of converting aural experience into inscribed fact.

Like phonography (and like phonotypy, palaeotype, glossotype, sound-hand, sound writing, and speaking type), the IPA and its antecedents all sought to broker similar transactions between the lasting, material expression of language and its experience as an ephemeral communicative medium, between the visibility and stability of texts and the multifarious sounds of speech. Each shorthand alphabet, spelling reform, or international phonetic alphabet presupposed that signs could represent sounds, yet each system also became embroiled within possibly damning competition or improvements. Each inventor somehow sought to be less arbitrary and less artificial than others, relying upon "scientific" principles, broad experience, or the wishful coherence and completeness of his alphabetic characters. Inculcated in each new system was an uneasy sense of bodily imperfection, ears not sure enough and hands not quick enough, and a related anxiety about the efficacy of memory, about experience slipping away, potential evidence squandered, and alien experience unaccounted for. Shorthand manuals and explorers' vocabularies cffered themselves as solutions to these imagined problems—regulating experience by making it textual, running it down and pressing it onto the page. In doing so they relied upon rationalist assumptions about the reach of science and the objectivity of representation, about the certainty of observation and the value of paperwork, fundamental to their own cultural identity, to Anglo-American self-possession, its bureaucratic means and imperialist tendencies.

Making history, whether making public memory of legislative debates and judicial proceedings or making public knowledge of Samoa, depended upon inscribed representations for which objective status might be assumed. On the evidence of phonetic shorthand and its nineteenth-century contexts, objectivity had two essential ingredients and one proof: accuracy and impartiality made representations objective; *reproduction* proved the point. Accuracy and impartiality arose in the nonarbitrary, scientific pretensions of shorthand alphabets, in the professional training and deportment of shorthand reporters, in the unthinking, automatic, or mechanical work of reporting, and the implacable authorlessness of reports. Authorlessness hinted that, like Fox Talbot's early photographs, shorthand reports were produced by the "pencil of nature" writing on a stenographer's notepad. Fox Talbot had in mind the rays of light hitting photosensitive paper, while stenographers had in mind the sounds of English streaming onto the page in perfect phonetic script. The authorless origin of Talbot's calotypes might be Nature, but the authorless origin of shorthand reports turned out to be something variously called *government* and *law*. The self-effaced, automatic person of the shorthand reporter occupies a liminal zone between authority and text that I discuss in Chapter 5. For the present, let me return to the invention of Edison's phonograph with a word or two about reproduction. As the missionaries in Samoa learned the hard way, reproduction offered one proof that phonetic inscriptions were properly representations of aural experiences. When the missionaries made oral mistakes "which appear[ed] absurd" to the Samoans, it was a sure bet that their impromptu phonetic alphabet was useless as a representation of Samoan speech. Shorthand inventors feared the same lesson as they shrilly promoted accedence to rules. Reporters feared it as they broadened the meaning of "verbatim," and legislators glimpsed it as they questioned reporters' methods or glowered over their own words in transcribed proceedings. If transcripts could be read to reproduce the very sounds that had been uttered in the past, then the reporters had done their job and their squiggling scripts had represented speech.

One final, eccentric shorthand alphabet demonstrates the extremity of contemporary desires for audible reproduction on the eve of Edison's phonograph, as well as the assumption—so soon controverted—that while such reproductions might be textual in their instrumentality, they were certainly human and anatomical in their performance. Alexander Melville Bell, father of the telephone's inventor, Alexander Graham Bell, published a shorthand alphabet called "Universal Line-Writing and Steno-Phonography; On the Basis of 'Visible Speech'" in 1869. It received little attention from contemporary observers, but testifies to the elder Bell's resourcefulness in using inscription as a means of ordering the world and regulating the body.[42] The "Steno-Phonography" pamphlet was a shorthand manual, published in a few rough pages the way many systems were, and was a sequel to the same author's "Visible Writing." To-gether the two works completed "for every purpose, the new science of Universal Alphabetics." The proclaimed universality of his sys-tem was extremely ambitious: "All speech-sounds are now provided with mutually related Symbols, adapted for Printing, for Writing, for Telegraphy, for enabling the Deaf and Dumb to Speak, and for teaching the Blind to Read." Bell's shorthand would work for civil and business reporting, but it would also replace the Roman alpha-bet and Morse code, and it would furthermore remediate the textual impairments of deafness and blindness. The key to all this "univer-sality" was a new version of "nonarbitrary" signs. Bell claimed to have invented symbols so scientific that they could reconcile a jum-ble of differently textual uses. His symbols were "mutually related" to speech in that each was based on a different vocal configuration of the mouth. In Steno-Phonography, for instance, the slope of a character indicates "the part of the mouth by which the sounds are formed." Alexander Jackson Graham had earlier used oral anatomy as a mnemonic map for some of his consonants, suggesting that his students imagine a face with the symbols for the labial sounds *p* and *b*, for instance, located near the lips. But Melville Bell actually based the shapes of his characters on the anatomical features and functions of speech. Thus was he confident that he could teach the deaf to speak and, by positioning actual mouths, the blind to read. Oral and aural experience did not need conversion into print in the same way as before, because steno-phonography was already an organic tran-scription—maybe even a textual reproduction—of bodily experi-ence. Speech equaled writing, or came a lot closer to being equal in Melville Bell's imagination than in the explicit aspirations of other authors. He injected the body into shorthand rather than projecting shorthand onto the body. Other systems had variously discerned fin-gers, eyes, ears, head, and hand, but Bell emphasized the mouth, in-scribing its movements on paper.

Despite Bell's extraordinary ambitions, the equivalence of short-hand alphabets and technology is not uncomplicated. Without qual-ification such equations prove too limiting of literacy practice, technology, or both. Shorthand involved material instruments, of course —the pen, the paper, and the printed manual—but shorthand was most keenly technological to the degree that it became located within the same rhetoric of progress, interrogation of subjectivity and anatomy, and battle for standards that so many new mechanical inventions were. Like those inventions, shorthand systems appealed to their times, signaling new desires, whether admitted, imagined, unconscious, or unfelt. And like more mechanical inventions, the success or failure of shorthand systems was predicated on more than the commonality or some retrospectively glimpsed worthiness of those desires; it needed a position within the economic and social realities, the perceptual and mental conditions of its moment, including the evolving public sphere, the bureaucratic organization of business and administration of government, and the shared experiences of textuality, its character and limitations.

This chapter has pursued the analogy between shorthand alphabets and machinery in order to explore issues of literacy, standardization, and reproduction, but the shorthand authors themselves drew the same analogy for different purposes. Shorthand was enmeshed within a rhetoric of progress that cast the reporter as a technician, contradictorily both skilled and automatic. It was a familiar, if often forgotten, problem of subjectivity, and it bubbled below the surface of the indeterminately defined "verbatim." Was the shorthand reporter the equivalent of a machine for "fastening thought upon paper," as one manual put it (Webster, xii)? Or was the reporter more like a pilot or an engineer, skillfully manipulating technology as he kept "his highest pressure of steam constantly on" (Woodward, 34)? Shorthand practice was rife with both assumptions, other metaphors emphasizing the materiality of language, and corresponding models of the writing hand, the speaking mouth, the listening ear, and the reading eye. And while shorthand authors interrogated the limits of textuality, related questions about the coincident limits of technology seemed to many individuals more immediate and more pressing. Shorthand boosters vaunted the conversion of experience into textual evidence and saw the reporter's body and alphabet as the necessary, structural instruments of that conversion. Inventors, science fiction writers, and other imaginative Americans meanwhile vaunted the potential of other, more patently technological conversion mechanisms. First among these was Thomas Edison's phonograph, to which the next chapter returns. Edison's tinfoil phonograph intervened into the complicated world of shorthand, in which sounds and mouths were variously and textually inscribed on paper, by hand, in alphabetic representations of speech, and in which reading texts was the one way to reproduce speech.

Imagining Language
Machines

While at L[ouisville during 1866–1867,] I got for the 1st time an insight as to how speeches were reported. The associated press had a short hand man traveling with Presdt Johnson when he made his celebrated swing arround [*sic*] the circle in a private train delivering speeches—The man engaged me to write out from his reading the notes. He came in loaded & on the verge of incoherence—we started in. . . . He would frequently change words, always to the improvement of the speech. I couldnt understand this & when I asked him that if he read from notes why these changes, "Sonny" said he, if these politicians had their speeches published as they delivered them a great many short hand men would be out of a job. The best short handers & the holders of good positions are those who can take a lot of rambling incoherent stuff &.make a rattling good speech out of it.

—Thomas Edison, Reminiscence, 1909[1]

The inventor Thomas Edison started out as a teenage telegrapher, tramping the Midwest and border states at the end of the Civil War. In April 1867, when he was twenty, he picked up a copy of Benn Pitman's *Manual of Phonography* in Louisville, where he was working for Western Union. Ten years later, after he had cut his teeth on telegraph circuits, the vote recorder, and telephone transmitter, he invented the phonograph. The device made him world famous and earned him notice in the press as "The Wizard of Menlo Park." It also initiated a flood of letters from the curious and the supplicant, from opportunists and would-be inventors, which Edison would receive throughout his long career. There is no evidence that Edison, who wrote in the quick, legible hand of a telegrapher, ever learned shorthand, yet the name he gave his phonograph indicates his debt to the climate of representation within which nineteenth-century shorthand developed and prospered. As Edison conceived it, the phonograph would be a business machine for the conversion of aural experience into *records*—permanent, portable, reproducible inscriptions. The aural experience he had foremost in mind was clearly speech, not music. The phonograph would be party to the textuality of American life, making text more mechanical and fulfilling its varied materiality as documentary evidence. It combined the phonetic interests of phonography with the certainty of machinery. Edison boasted of his invention in an article for the *North American Review*, which had a long tradition of publishing on the subject of language in America.[2] He itemized its use for talking clocks and dolls, and for taking dictation and recording novels.[3] The dead would be able to speak, the blind to read. Technology would reconcile experience and evidence in ways that shorthand had failed to do in any but legal proceedings.

From amid the immediate fanfare of news accounts, public exhibitions, and private audiences, there was ample evidence that the public uncritically accepted Edison's phonograph as the inventor envisioned it. The device was potent with read/write functions, not just the read-only function that the twentieth-century entertainment industry would later monopolize. Textuality and amusement were not as firmly opposite as they are taken to be today, but the phonograph tended decidedly toward the former. Language and elocution teachers seized upon it. Authors, ethnologists, and ethologists wrote to Edison hoping the device would ease their labors, while the people most worried about its implications were stenographers. Immediately in 1878, the New York State Law Stenographers' Association heard a paper on the matter at its annual meeting in Rochester. The paper was derisive, facetious, and revealed something of the culture of inscription where shorthand made sense and the phonograph made trouble. Its author spoke of trying the phonograph in court, only to be frustrated with the small amount the machine could impress on each cylindrical record. The necessity of reproducing the court's aural experiences in real time also proved frustrating. It took the same amount of time to listen to the proceedings a second time; therefore, checking a matter of evidence or transcribing the whole required different commitments of time than shorthand did. Next the author reported an accident by which he learned the phonograph could "stereotype" any object introduced into its "funnel." A cat in the funnel produced kittens; strawberries berried out of season; money reproduced itself. The stenographer's joke shows just how material aural experience was to the culture of shorthand. Sound was an object, like a cat or cash, and having been made material, it could provide an object for the phonograph, just as print or typeface provided the matter of stereotype printing.

Phonography and phonetics had disassembled language in order to study it. The German physicist Hermann von Helmholtz and others had analyzed and simulated vowel sounds, seeking the ingredients of sound and sense. Now the phonograph seemed to hold the promise of new accomplishments along similar lines. In particular, Edison struggled with sibilants, which did not record well on his original tinfoil recording surface. For the moment, sibilant consonants replaced vowels as the fundamental subject of acoustic and linguistic science. The machine began "to increase our knowledge of the parts of speech," wrote one observer to the *Telegraphic Journal and Electrical Review*.[4] The next decade confirmed the phonograph as a language machine. Even the music critic of the *New York Post* could not see what was coming; he commended Edison's improved machine of 1888 for its reading of *Nicholas Nickleby*, bubbled at the prospect of phonographic newspapers, and assumed the benefit of texts that could "read themselves," particularly for readers in remote or isolated (non-English-speaking) locales. As for music, its study would be improved. A composer's own interpretations and the fleeting genius of improvisation would both be recorded for posterity. The music critic's daily reviews could even include samples of the music he discussed, once newspapers had been replaced by wax records.[5] He was serious. The first glimpse of a paperless world and of its multimedia hypertexts was exhilarating. Paperlessness was a way to keep things live, to save the vibrance and authenticity of experience without succumbing to the dryness of textual evidence and the arcania, or obscurity, of different notational systems.

The point was elaborated by Edward Bellamy, who, coasting on the rising sales of his *Looking Backward* (1888), published a utopian science fiction about the phonograph in *Harpers* (1889). In "With the Eyes Shut" Bellamy gives another framed narrative in which a dreamer imagines the future. There the phonograph has taken over most of the discursive functions of culture. Instead of reading, people are read to: they can read with their "eyes shut." Phonographs are called "indispensibles," fulfilling the roles of newspapers, books, shopping lists, sermons, memo pads, and more. In his evocation of a different future, Bellamy necessarily implies much about the present; his story consequently indicates much about the act of reading in 1889, at the same time that it forecasts a future in which textuality—ubiquitous and in some respects homogenized— has been transformed by a new technological medium. Notably, Bellamy's fictional critique of reading resembles phonographers' criticisms of "common" writing or shorthand authors' condemnation of one another's systems. In his enthusiasm for the phonograph, Bellamy's narrator decries the "roundabout means of spelling out the signs that stand for words, and imagining them uttered, and then imagining what they would mean if uttered" (737). Here are the phonographers' assumptions that utterance is the root of all writing and that arbitrary signs are its bane: artificial, unnatural, irrational, and imperfect. Reading and listening to the phonograph are both processes of "absorption" in Bellamy's story, but the "necessary fixity of the eyes" that real reading requires makes the reader less "responsive" to the matter being absorbed. The narrator watches the import of phonographic texts play across the features of their listeners: phonographs emotionally liberate readers. They also bodily liberate them: posture is better and eyesight improved. In becoming less arbitrary and less arduous, the phonographic "reading" becomes more natural and somehow more human.

Bellamy's narrator further dreams that writing is approaching a lost art, that written English will become a dead language. The idea of an obsolescent or moribund English was neither unique to fiction, nor particular to technological causes. Linda Dowling identifies such an anxiety in the contemporary field of comparative philology, tracing its impact upon decadence, that late-century agglomeration of aestheticist movements dependent upon the "portrayal of written language as an artificial and usurping power" (175). Two early hallmarks of the decadent tradition were Walter Pater's assertion that all art in its perfection tends toward music, and Algernon Swinburne's tonal aspirations, his onomatopoeic voice.[6] As coincidental and as mutual as the aural sympathies between decadence and shorthand may have been, the phonograph seemed to affirm them both in this regard. Edison's device may have been intended for text, but it made writing more problematic, painting it arbitrary and recherché. It also promised new kinds of meaning, less textual texts, and fresh possibilities for style, genre, realism, and voice.

Though certainly with less method and erudition than decadence, a much broader aestheticism was emerging. Apart from literary circles and outside the reach of Pater or Swinburne, an unconsidered, amoral appetite developed around the phonograph and similar devices. This was another decadence, a new culture of mechanical amusement, which had previously been limited to the social elite and now became an object of middle- and working-class desires. Scholars have noticed the origins of the emergent culture of mechanical amusement in the modest rise of disposable incomes, particularly among single, urban women, as well as in the growth of urban centers, in the related pressures to the definition of private and public spheres, and in a febrile ideology of American abundance and commercial prosperity.[7] Before the stunning rise of the nickelodeon around 1908, one of the more obvious registers of the new order was the nineteenth-century shift in focus from the Victorian crystal palace to international expositions like the ones at Paris and Chicago. Though no less exoticist and no less devoted to demonstrating Western "progress" than London's Crystal Palace (1851), the World's Columbian Exposition in Chicago (1893) offset its gleaming "White City" with a carnivalesque "Midway," where displays of American progress gave way to more prurient interests. This new institutionalized dialectic between white city and midway burgeoned; everywhere midways appeared by themselves, at the end of streetcar lines, at seaside resorts, and as the outgrowths of Barnumesque "museums." Amusement parks and arcades willfully divorced mechanical amusement and commercialized spectacle from the instructive achievements of the white city. They liberated novelty and self-indulgence from the double baggage of self-examination and self-congratulation.

Ample evidence of the change emerged unwanted at the annual conventions of the first phonograph distribution companies, which revealed just how explicitly antipragmatic modern amusements were. Attending capitalists seemed all but oblivious to the rising tide of mechanical amusement, so sure were they of a utilitarian basis to trade. Their misapprehensions prove instructive. The so-called "local companies" early on acquired territorial rights to Edison's phonograph and its competitor, the graphophone. By agreement these local companies sought to lease phonographs and graphophones to business offices within their respective territories. Leasing, rather than selling, followed the model of telephone distribution and allowed corporate control over individual machines, which continued to require periodic adjustment and repair. Dictation remained the unquestioned, utilitarian purpose of the phonograph/graphophone, though there was much hope that the wax record would soon replace the business letter, as executives became convinced to send and swap recorded cylinders. The genre of the memo was still being established;[8] perhaps it would settle on this new form. The First Annual Convention of Local Phonograph Companies was held in Chicago in the spring of 1890. The title page to its *Proceedings* announces that stenographers were not hired: sessions were "Reported by the phonograph and phonograph-graphophone." Not surprisingly, attending capitalists complained that the new machines were meeting resistance from office workers. This resistance was particularly keen with the "amanuenses," and what one participant disparaged as the "nickel-in-the-slot stenographers"—in today's parlance these are the secretarial aides and office temps. By contrast, "first-class" stenographers apparently knew progress when they saw it; some even leased their own phonographs. The Columbia Phonograph Company, whose executives were themselves former stenographers and whose territory included Washington, D.C., reported contentedly that "some sixty machines" had been placed in the offices of congressmen (57, 53). Elsewhere there was disquietude in the ranks. A phonograph agent from Dallas complained that businessmen declined to lease machines if they encountered any "musical exhibition." The potential customers went away marveling, impressed, and yet they never seemed to place an order. In fact, the agent reported, "We have left positive orders that no one is to hear music on the phonograph at the business office; to give a business exhibition and nothing more" (72). Amusement seemed to contaminate the pragmatic purpose of the phonograph. Participants to the convention wanted the two functions to be entirely separate, because they considered amusement so wholly secondary to their product and business.

To be sure, part of what was going on was the confusion of businesslike means with businesslike ends. The phonograph executives mistook their own interests for the interests of consumers. Accordingly, they assumed that low-paid office staff and good-time Charlies lacked the necessary skill, attention, and incentive to operate the phonograph correctly or maintain its still-quirky mechanism. Participants at the convention listened with fascination—a curiosity not unmixed with horror—as they heard one entertainment pioneer from California describe nickel-in-the-slot phonographs, amusement machines strategically located in saloons, depots, and other urban crossroads. Nickels had come pouring in. Subsequent trails showed that this was no mere fluke: during the next year, hundreds—in some versions thousands—of dollars were earned per month by a machine or two, nickel by nickel, penny by penny. Like the lamented recalcitrance of nickel-in-the-slot stenographers, the eruptive popularity of nickel-in-the-slot phonographs proved to be a wedge that opened the modern entertainment market. Though local companies were soon bankrupt or dispossessed, the phonograph surged. Consumers had collectively changed the market and appropriated the phonograph. The machine's new consumers were, in a sense, much more passive than its previous ones, because they listened to prerecorded musical selections rather than recording on their own (satisfied with ROM instead of RAM, to use a contemporary idiom). Simultaneous with the satisfactions of this "decay" in "aura," as Walter Benjamin puts it, the social context of the phonograph started along a trajectory from public parlor to private home, its market profile from arcade novelty to meaningful domestic possession. Prerecorded musical selections distanced listeners from their powers of origination as recording artists, while the personal, domestic comfort of the listening act itself distanced their amusement from the kind of public life that is experienced in public space.

In two following chapters I will address different features of the amusement machine, its timely provocation of questions about authorship and reading as well as its problematic, double construction as an objective medium and popular diversion. In different ways, both the phonograph and motion pictures heeded and challenged the climate of representation that characterized the final years of the nineteenth century and the first years of the twentieth. The outcome of such give-and-take was twentieth-century popular culture, a new phenomenology and pattern of commerce that still inform mechanical amusements. Before drawing such connections, however, this chapter continues the articulation of that ambient climate of representation, which fostered phonography and expected the dictaphone. The inventor of the phonograph, contemporary pundits, novelists, and capitalists had all misconceived the phonograph. They had all been wrong in pronouncing the function and the future of the new technology. This chapter forges along that telling path of misconceptions, or, more properly, of preconceptions. Though quickly swept up in a changing entertainment economy, the phonograph and similar inscriptive devices continued to provoke potent senses of psychological and cultural orientation that had little to do with amusement and were no less suggestive of American textuality than the coalescence of material(ized) authority in the shorthand practices of the Reconstruction years. If the authorless texts, arbitrary signs, and "verbatim" objectivity of shorthand suggest contrary experiences of texts as both eminent representations and strangely inadequate reproductions, then the mistakes and preconceptions surrounding late-century inscriptive technology indicate a similar range of coextensive possibilities.

~

Whether they possessed technological expertise or not, Americans struggled to orient themselves within a world and climate of representation that they understood as importantly technological. Whether called modern or modernist, their new orientation involved parameters of time, space, economics, and aesthetics, as Stephen Kern, David Harvey, and others have so ably demonstrated. The sources of such orientation included new technologies and practices of communication, transportation, and inscription, as well as a powerful sense of technological potentiality, of what might soon or had already been achieved. Feeding this sense of potentiality were the assumptions, preconceived needs, and desires that would help make some future inventions successful and others unsuccessful. Self-consciousness about progress, manifest in a shared sense of national or cultural purpose that had already helped elevate Edison, Bell, Tesla, and other inventors to the status of almost cultlike celebrity, further fed this sense of potentiality. In pursuing questions as nebulous as a climate of representation and a sense of orientation, I have relied upon a variety of sources—fact and fiction—and particularly upon a group of long-neglected expressions of technological potentiality, of intention rather than invention. Edison's celebrity made him a lightening rod for the schemes of others, and he received letters by the thousands from erstwhile inventors seeking advice and encouragement. The "idea letters," as they were called, provide glimpses of what seemed likely or possible at the time and demonstrate the terms within which new technology and practices of representation were conceived. They show, for instance, that inventions like the phonograph, notwithstanding the culture of mechanical amusement, continued to offer a partly textual point of reference that helped enroll individuals within new and sometimes contradictory senses of the available literacy practices.

As my narrative of the dictation phonograph demonstrates, sources for the study of technological preconceptions are not always hard to come by, but they can be difficult to evaluate. One participant at the local phonograph company convention, for instance, reported that a visitor to his office "felt quite disappointed upon discovering that the phonograph would not answer their questions. 'Why,' they said, 'we thought you could talk to it, and it would answer your questions'" (33). The report was likely untrue, a joke of the sort that helped bond convention goers into a single community of experts. But it is hard to tell. Similarly, when phonograph users reported playing records backward and hearing word order reversed, or when Bellamy's fiction included a phonograph that could speed its record way up and still be intelligible, their sources of error are hard to identify (in this age long before "Paul is dead" and The Chipmunks). Were the assumed textual purposes of the machine and the inevitable analogy to printed books simply strong enough to occlude the facts? Or were Bellamy and others adding some literary license to their reports of the phonograph's luster? The same revealing problems with sources continue even in the realm of technological expertise. In 1888, as Edison was putting the finishing touches on his new laboratory complex in West Orange, New Jersey, the inventor started an experimental notebook he labeled his "Private Idea Book." In it he listed 111 things he planned to invent. Within two years he had begun another, distinctly different representation of future technology, a science fiction novel. Though Edison abandoned both books after filling only a handful of pages, his efforts indicate the irregular shape and character of technological potentiality as it appeared to the most celebrated expert of the period. Intuitively, the preconceptions of the idea book should be easier to read than those of the novel, but fact and fiction seem to break down as the two books glimpse technological worlds that were plausible in some sense, within their own contexts, to the man who knew so much about extant technology.

Thirteen of the inventions Edison listed in his idea book have to do with phonographs, including applications like "large phonograph for novels," "toy phonograph for dolls," and "phonograph clocks." At least thirty-six more have to do with communication or inscription—telephones, telegraphs, photographs, mimeographs, meters, and railroad signals. This number does not include the many batteries, dynamos, and motors that Edison may have intended to use with such devices. In some of Edison's ideas it is possible to glimpse his sense of both mechanical amusement and print culture. He planned to invent an "electrical piano" as well as a process to "soften the ink of books," transfer it to copper sheets, and produce electrotype or stereotype plates—a publisher's nightmare and a copyright pirate's dream. The list also includes some tantalizing conundrums, like "ink for the blind," which attest to Edison's interest in representational practices, but which further suggest an indeterminacy or a textualizing plasticity similarly apparent in earlier developments, like Melville Bell's visible speech and Isaac Pitman's phonography. Whatever Edison meant by it, "ink for the blind" challenges the norms of perception and offers a new, textual materiality. Nor was text the only new material Edison wanted to test perceptions with: the idea book also harbors the promise of new commodities, such as artificial silk, ivory, hard rubber, and mother of pearl. Edison was imagining plastics twenty years before any such things existed.

The inventor's science fiction notes of the early 1890s, like his idea book of 1888, list rather than narrate. Edison's tentative title was "Progress," and his notes give half sentences and short paragraphs that summarily leave matters of plot and character to his collaborator, the writer George Parsons Lathrop (son-in-law of Nathaniel Hawthorne and associate of Mark Twain). Both men wanted to cash in on the success of *Looking Backward*, so there is a Bel-

lamy-like tone to Edison's notes, which give a fragmentary ethnography of a future society, dressed up with a little astro-geologic fantasy, including antarctic volcanoes, planetary collisions, and a shift
in the earth's rotational axis provoked by the Suez Canal. Like the
idea book, the science fiction notes display a striking prescience—
Edison imagines infrared photography, for instance, as well as nylon
and plastic. But they also bare the latest twist in Edison's preconceptions of the phonograph and amusement. In "Progress," the
phonograph and Edison's kinetoscope (a precursor to projected motion pictures) have been combined to reproduce dramatic and oper-
atic performances for consumption at home. The result has been a
withering away of legitimate theater and, in Edison's shorthand
phraseology, "Every family wealth." Although he has admitted their
use for amusement purposes and forecasts so much about the TV
and VCR, Edison still sees mechanical amusement as a means of social uplift. The new media will democratize drama and opera, allowing access to these improving arts for consumers otherwise dis-
advantaged by economics or geographical distance from the theater.
Utilitarianism still underlies Edison's invention, though here it is
part of a broader social program glimpsed elsewhere in the science
fiction notes as chilling eugenics laws and police surveillance by hot-
air balloon.

Again like the idea book, the science fiction notes prove most in-
triguing when they are most opaque, when they degenerate into a
list of curiosities, like "hypnotizing machine," "7th sense," and,
particularly, "color music." Did "7th sense" mean that Edison had
accepted a sixth? The putative distinction between nonfiction and
fiction, plan and fantasy, becomes nearly negligible when considering Edison's "ink for the blind" beside things like his "color music."
Both collapse visuality into aurality. The question becomes not only
"What did he mean?" but further "What could he have meant?"
These are conditions and questions that apply equally to the "idea
letters" Edison received throughout his career. The idea letters oc-
cupy a position that can be called neither fact nor fiction. Some of
them propose ideas that are ridiculous, others practical. What they

share is a species of intention; they are an accidental genre.[9] Like Edison taking his notes or trying his novel, the individuals who wrote to the inventor did so specifically, directly, acting within an epistolary genre they found self-evident in character. They were expressing technological possibilities as decisively tailored to the conditions of their composition and expectations of their readership as either of Edison's aborted books. By proposing answers, the idea letters and Edison's books demonstrate the questions that technology was assumed to address. The answers are necessarily speculative, constrained by a variety of circumstances, and can retrospectively seem implausible or remain opaque. But the questions are what matter most. Because multivocal, the letters are a particularly powerful means with which to identify the questions of the era. Frequently suggested ideas, like the discursive habits of their suggestion, mark the kinds and aspects of technology that remained unsettled and possessed weight precisely in their status as open questions. Plurality, in this case, acts as a tonic for anecdotalism.

Like the correspondence files amassed by similarly public figures,[10] the collection of letters housed at the Edison National Historic Site in New Jersey can be divided topically and thematically in order to reveal themes inherent to the mass. Documents at the Historic Site archive number more than five million. Of these, unsolicited letters to Edison probably constitute more than 100,000 items, including fan mail, personal requests, and inquiries, as well as the idea letters. Approximately 20,000 unsolicited letters from the period 1888 to 1915 have been examined for the present study. Idea letters make up roughly half the total. Collectively, they reveal patterns of expression that vouch for their value as a source; they also illuminate Edison's stature as a cultural icon, the process of invention as it was popularly understood, and the ongoing negotiation of boundaries assumed for technological function and expertise. In addition, an important subset of the idea letters has been selected for further examination: this smaller group of letters has as their subject inventions that apply technological solutions to perceived needs in the areas of inscription and representation. They propose printing machines, phonographs, registers and signals of many kinds. In doing so they compliment Edison's interests while confirming what recent scholarship has identified as a central preoccupation of their age.[11] They second Edison's idea book and confirm the interests of the phonograph convention. They imagine language machines.

The people who wrote to Edison may have collectively formed part of a culture, or cultures, but each was also idiosyncratic in his or her circumstance and expressions. What they had in common was that each recognized Edison as a person with whom to communicate, whether to complain about his products, beg an autograph, garner advice, or propose an invention. They did not constitute a textual community in the sense that each wrote in isolation to the man he or she considered an expert, an insider in a perceived community of men who did technological work and acted as spokesmen, interpreters, and promoters of technological change. If they wrote to Edison as insiders, then they considered themselves outsiders to some degree, yet their "outside" was never so distant from Edison that they did not expect or hope for a response, advice, or assistance. In this they formed an unconscious culture of letters,[12] a folk culture in which individuals asserted their connectedness to the trends and machinations of modern life. Each letter was a recognition of the authority of both technology and Edison. The gulf between expert and amateur was far greater from Edison's vantage point; most of these unsolicited letters were dismissed with a perfunctory reply or no reply at all. The selective permeability of the insider-outsider frontier bears emphasis. Part of Edison's stature as a cultural hero stemmed from perceptions of his own rags-to-riches crossing of the same divide. In fact, he remained an insider to his correspondents long after he ceased to be one. Well into Edison's sixties, seventies, and eighties, correspondents wrote regarding technical matters he had long since abandoned or forgotten. Some continued addressing their missives to Menlo Park, decades after the inventor had removed to West Orange.[13]

The yawning gulf between the authors of these letters and the proprietors of technological knowledge of the period is obvious to modern readers, as it was obvious to Edison and his associates. Yet the full vastness of the gulf was clearly not part of the experience of the letter writers, whose visions of technological possibility were just that—possible—to them at the time they wrote. The history of technology has until recently been a teleological discipline, however, and even the best cultural historians of technology can seem to elide the point. As David Nye explains of the early "social meaning" of electricity:

There was never a time when ordinary Americans understood electricity in purely functional terms [. . .] Indeed, "impractical" dramatic lighting in the theater predated Edison's "practical electric light," and until after 1900 the average person was much more likely to have contact with electrical medicine than with home appliances. (*Electrifying*, 382)

The implication that electrical medicine (electric belts, "vitalizers," and tonics of various sorts) was impractical to the people who used it, or removed from a functionalist model of electrical process, warrants correction. Electrical medicine was used exactly because it was considered salutary, and because electricity and metabolism were understood to function congruently, however mistaken the specifics of this understanding later proved to be. In fact, popular interest in technology has often been functionalist in the truest sense of the word: the idea letters are about machines or devices intended to do things in particular ways, whether they later did them or not. Letter writers reveal their investment in the question of "how things work," with the same directness of Edison compiling his laboratory notebooks. Answers to the question varied greatly, of course, and historians of technology are now beginning to pursue "failures" and alternatives as a subject of inquiry.[14]

Resisting the logic of teleology can be difficult; resisting its rhetoric can seem impossible, not in the least because a few of the ideas proposed by Edison's correspondents seem so absurd. Letters arrived regularly asking about the "Edison Star," which had been a newspaper hoax—an electrical star that Edison reportedly launched from Menlo Park. The files of letters received in the 1880s as well as the trickle of queries that continued well into the twentieth century indicate something about the way electricity was experienced during these years. A similar stream of letters proposed various perpetual-motion machines, which suggests the durability of an idea long discarded by experts. When individuals tinkered with perpetual-motion machines, they demonstrated a tellingly outdated understanding of energy and machinery. They experienced energy as a valuable, produced commodity, and machinery as a versatile fact of modern life. That these realities seem to have overridden the finer points of physics indicates the unquestioned weight such understandings had. The surprising persistence of the Edison Star and perpetual-motion machines points toward a technological folklore, a word-of-mouth culture of technological possibility every bit as tenacious as other folkloric traditions.[15]

The "idea letters" received by Edison are a nebulous lot, making a definitive, quantitative sample of them impossible, if only because their identification was—and remains—somewhat arbitrary. (In the current archival arrangement, perpetual-motion proposals count, Edison Star inquiries do not.) The extant archive does allow of some general impressions, however. Idea letters came from everywhere, but mostly from the United States, from both rural and urban areas. They came from women and children, though the majority were from men. They came from grocery clerks and housewives, medical doctors and farm hands. Writers usually specified lack of training or lack of capital (or both) as their particular claim to outsider status and their impetus from writing. Sometimes many writers would seize the same occasion to write, a news story carried on the wire, for instance, like one incorrect report of 1908 that Edison was working on aerial navigation. In such cases hundreds of correspondents seemed to feel that their ideas had been solicited by the inventor—as they really were during World War I when Edison and other members of the blue-ribbon Naval Consulting Board made a public appeal for suggestions. During and after 1917 thousands of letters poured in with ideas for weaponry and other war *matériel*. Familiarity with the archive, its provenance and organization, indicates the likelihood that some idea letters were removed or distributed to other files: some were almost certainly discarded; others smacking of business correspondence, e.g., those on commercial letterhead or invoking commercial references or concerns, were more likely routed elsewhere. Despite this sorting, the remaining idea letters are far from uniform. Typescripts on printed letterhead lie beside handwritten notes on crumpled loose-leaf; the grammatical and well appointed lie beside the unlettered and illegible.

Despite their diversity, the idea letters do exhibit identifiable patterns of expression. The most surprising thing about them is that so many concern ideas, not inventions or discoveries, not machines or devices. "Are you in the market for ideas?" queried a livestock merchant from Kansas City, "I have one that I believe would prove serviceable" (Waters 1905).[16] Like the word "curiosity" during the previous century, the word "idea" here denotes an epistemological currency of learning, a unit of knowing that connects an individual's mind to an experience of nature. But unlike the curiosity cabinet, the idea always possesses a kind of exchange value in the sense that people *have* ideas, and *having* appears to have been far easier than inventing. Two days after the livestock merchant wrote, another man inquired, "Do you ever help invent another person's idea?" (McTillen 1905). He writes as if inventing and having an idea were entirely distinct: ideas involve possession while inventions involve action. Writers beseech Edison to "work out" their ideas; the word "out" is as operative as the term "work." "I have several good ideas in my head," writes one man, "but that is as far as they ever got" (Braymer 1915). Ideas are written out, worked out, tried out, carried out, perfected, and made practical. In their outward progress, their path away from the self, they are invented. Their trajectory carries them toward a series of imagined marketplaces, one where ideas are bought and sold, and another where inventions may prove successful. The goal of many correspondents was to become what one termed "a multi-million air" (Parliman 1905). They were local utopianists, envisioning the betterment of their conditions as well as the benefits to consumers that might result from their inventions.[17] Both their focus on ideas and their characterization of invention as a complicated form of self-expression offer a necessary context for the title of Edison's "Private Idea Book," which must have derived from similar impulses.

Writers frequently capitalize the noun "Idea," sometimes enclosing it within quotation marks. Their communications are idealist in a very literal, self-conscious sense. More than a few write from an orally or phonetically centered universe in which the word is spelled I-d-e-a-r, suggesting a folk culture in which people have ideas, talk about them, but do not read about them with any great regularity. Even the "crank" letters, which Edison's secretarial staff frequently distinguished from the idea letters, had plenty to do with ideas per se. A man named Ike Leonard Isacson wrote in 1914:

Dear Sir,

    I am a patient, in the Hospital for Insane, at Elgin. Illinois. I am not Systematized. The Idea is. I must be a System. I wish you will please Install a System for me. Hoping to hear from you. I remain. Very respectfully,

                                Ike Leonard Isacson.

His handwriting is steady, neat, his plight pathetic. Across the top of his postcard Edison has written his familiar "Noans," an abbreviation he used to specify "no answer." Whether with understanding or without, Isacson puts his finger on the crux of matters. There is no way to know whether and in what proportions Isacson's desire for a "system" arose organically from his own mind, or was planted in him by the Hospital for Insane, medical discourse, and the normative (i.e., systemic) conditions of society. His postcard presents a riddle rich in significance for all of the idea letters, which each arrived with its author's hopes to "get systematized" in some different yet specific way. Isacson's simple sentence, "The Idea is," appears with surprising frequency in other letters.

Even some of the noninstitutionalized letter writers might be called possessive compulsive. They represent ideas as unavoidable impulses, obsessively retained. One man from Cambridge, Massachusetts, complains, "It is a great problem to me to find an outlet for ideas which I am continuously working out." Another man

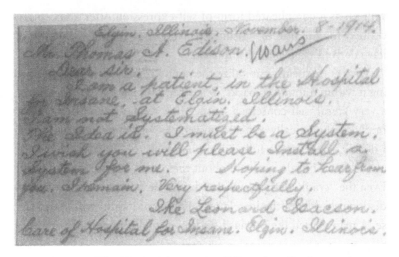

FIGURE 4. "The Idea Is." Postcard to Edison from Ike Isacson (1914).

writes with greater self-consciousness, "I have been a victim of the perpetual motion theory for several years" (Osgood 1913; Marsh 1914). In such cases the compulsion to have ideas translates into a compulsion to write to Edison as well as a compulsion to invent. Writers seem isolated by their ideas as well as strongly self-identified with them. Having ideas seems a chronic condition for many. Hundreds write that they have been brooding over an idea "for some time," or that an idea struck them "some years ago." The very durability of their ideas, like the longevity of the Edison Star or the persistence of Edison as a recognized authority, speaks to a sense of continuity indifferently at odds with the novelty writers' claim for their ideas or the tumult of technological newness hailed by the culture at large. Another form of the possessive compulsive is the writers who have ideas but who feel that they cannot divulge their secret or, more commonly, feel they cannot divulge their secret to anyone but Edison, for fear of having it stolen. Correspondents sometimes write many pages without getting to their ideas, finally indicating that a second letter or an interview will be required for them to divulge their secrets. In such cases the idea letter is only about possessing an idea, only about approaching Edison with its secret.

These retentive writers are seconded by the ones who complain that Edison *has* stolen their ideas. A factory worker from New Haven writes that "About five years ago" he had proposed to Edison a combined motion picture and phonograph machine, something Edison had in fact clearly contemplated as early as his science fiction notes. The writer explains that his proposal had been rebuffed by Edison at the time:

> Mr. Edison was so old he did not care or wish to take up any thing new in an inventive or commercial way. *Now* low and behold Mr. Edison comes out with exactly the *idea* I advanced to him and was turned down and I shall always feel that his idea sprung from my ideas sent to him. (Beecher 1913; emphasis in original)

Such expressions reinforce the fact that writers perceived ideas as property, private and personal, with little sense of collectivity and no recognition of the inevitability avowed by so many contemporary paeans to progress. "The idea" constituted a unit of understanding associated with technological problem solving, its value assumed to be a convertible currency, able to make the progress from the psyche of an isolated owner-citizen to the public world of technological elites and institutions. A republican ideal to be sure, fraught with the ideology of the American dream: as one self-identified "poor little country raised Texas woman" wrote from Oklahoma, her mind— the valuable idea it harbored—was "as undeveloped as her native lands" (Dale 1915). Technological possibility was anybody's grubstake, and letters to Edison, like letters to the editor, were part of the arsenal of the public sphere in America, ready-to-hand instruments of republican participation, gesturing at an abstract, rather than personal, level of connectedness.

～

Because the idea letters arrived from such different and widely dispersed individuals they index the cultural saturation of technological potentiality. If a small group of them evidence a tenacious folk tradition of perpetual-motion machines and other "possibilities" at odds with the knowledge of contemporary technological and intellectual elites, then the majority attest to the pervasiveness of what can only now be identified as more mainstream concerns, one of which is the rapprochement between technology and textuality. Like the authors and users of shorthand, like Edison and his early phonograph agents, the letter writers were absorbed in matters of message transmission and inscription. Many and various language machines seemed immanent. Persistent proposals for new, improved means of reproducing, inscribing, and communicating suggest representation was the site of anxiety, an open question that the letter writers were striving to answer even as they penned the time-worn phrases of polite epistolary discourse. Their ideas attest to a heterogeneous and unselfconscious sense of inevitability: the varied practices of textual and graphic representation constituted a frontier upon which the progress of American technology was both manifest and manifestly destined.

Among the frequently suggested ideas were many involving the phonograph. Writers proposed new uses for phonographs, improvements in Edison's mechanism, and new combinations of the phonograph and other devices. Notwithstanding the ever-increasing sway of its amusement purpose, the phonograph remained a language machine for many letter writers. One idea that repeatedly found its way to Edison's mailbox was the phonograph-clock. Beginning after his much hailed invention of the phonograph in 1877 and his widely publicized "improvement" of the device eleven years later, individuals wrote to Edison with the idea of a clock that would call the hours instead of striking them. One man wrote, for example, "My mind has been impressed for some time with the idea of a clock that would speak the time" (Norris 1915). Letters proposing the same invention kept on arriving into the twentieth century and continued until as late as 1920. Edison himself had clearly had the idea as early as November 1877, before he even got the phonograph working and included it in his published description of the phonograph's potential in 1878, entitled "The Phonograph and Its Future." He had listed it again in his "Private Idea Book." To letters received around 1908, he took to responding that the idea had been carried out in Europe, but that there had proved no market for the invention. Whether unbeknownst to the letter writers, or simply unacknowledged by them, the phonograph-clock was part of their cultural milieu. The continued recurrence of the phonograph-clock as a "new" idea confirms that the cultural saturation of technological knowledge was a matter of preconscious as well as conscious mentality. That is, many people came up with the same thing at the same time because the idea of the phonograph-clock percolated within the ambient culture. More importantly, they kept coming up with it because it kept percolating, a fact that suggests the phonograph-clock possessed certain symbolic features that pertained broadly to individuals' experience of public and private spheres at the end of the nineteenth century and the beginning of the twentieth.

What were those features? The clock is by itself among the most richly symbolic devices; Lewis Mumford calls it the "key-machine of the modern industrial age." Clock towers "almost defined urban existence" (14).[18] Preindustrial church towers meted out parameters of social identity and psychological composure: in the Old World, London Cockneys defined themselves as a group born within earshot of the bells at Bow Church. Marcel Proust made the receding steeples of Martinville one cynosure amid his *A la recherche du temps perdu*. Inside the home, clocks introduced an urban current. This held true even in rural areas, where they were sold by urban-identified salesmen or purchased from urban-based mail-order houses. During the nineteenth century, the clock maker was something of a "type" in the New World imagination, embodied by Thomas Chandler Haliburton's humorous character, Sam Slick, an itinerant clock seller true to his name. For all, clocks regulated private lives and connected their regulation to the patterning of social activity elsewhere. They mediated between the private and public, the individual and collective. More pointed forms of the same mediation transformed the workplace in the nineteenth and the early twentieth centuries. Factory time clocks, time-motion studies, and assembly-line time each added new weight to the symbolic burdens of the clock.

The comparable meanings of the phonograph are harder to parse out. Wyn Wachhorst has tried to describe the cognitive jolt the phonograph must have produced when Edison's invention was first unveiled. The very idea of a "talking machine" seemed impossible, the term an oxymoron. It denoted a contradictory combination of biological and mechanical function, a nineteenth-century cyborg. Further, this shocking mechanical paradox possessed the unique ability, as Edison had put it in 1878, for "the gathering up and retaining of sounds hitherto fugitive, and their reproduction at will" (527). What once was lost could now be kept. Fleeting, "fugitive" words and noises could now be captured and preserved. Another observer put it this way: "The phonograph renders it possible for us to seal the living speech on brazen tablets, and store it up."[19] Using the phonograph moved together sharply and decisively the two discursive grounds of speech and writing. Speech became differently ephemeral because it could now be saved, "stored up," and "reproduced." Shorthand experts had made verbatim transcripts into matters of skill and record; the phonograph seemed to turn speech into record and fact.

Individuals who wrote to Edison proposing the phonograph-clock assured him that it would be popular, but they did not specify their reasons for thinking so. While it is easy in hindsight to argue that the phonograph-clock would not have become widely popular (for the very reason that it didn't), it is more difficult to explain why the phonograph-clock persisted within the folk culture of technological potentiality experienced by these letter writers. Likely the device combined or contrasted the symbolic meanings of its two components in compelling ways. The phonograph-clock might have challenged the unrelenting progress of the clock with the phonograph's own ability to record fugitive sounds, to capture *temps perdu*, balancing lost time against leisure time. Or the phonograph-clock might have personalized, or humanized, the experience of time that clocks alone, in their persistent evocation of public life, rendered increasingly social and communal instead of private and individual. Probably the appeal of the phonograph-clock had something to do with relationships necessarily negotiated between private desires and the regularity of social activity. Both Edison in his *North American Review* article and Edward Bellamy in "With the Eyes

Shut" implied as much. Edison proposed the phonograph-clock, assuring his readers "The phonograph clock will tell you the hour of the day; call you to lunch; send your lover home at ten, etc." (534). In Bellamy's short story, a gentleman traveler is awakened abruptly in his hotel bed by a young woman's voice saying, "My Dear Sir, you may possibly be interested in knowing that it now wants just a quarter of three." The effect of these words is so "thrilling and lifelike" that Bellamy's modest protagonist gets up and gets dressed before he lights the lamp to investigate. He spends the rest of the night lying awake, "enjoying the society of [his] bodiless companion and the delicious shock of her quarter-hourly remarks" (737). In both Bellamy's fiction and Edison's wry prognostication, the phonograph-clock speaks with the voice of decorum, yet for both authors that decorum exits in opposition to titillating circumstance.[20] For many others the phonograph-clock must have inhabited a similar symbolic terrain, interrogating the familiar clock, its regulation of private desires and patterning of social activity. And maybe it was the discomfort of this interrogation that kept the phonograph-clock and its "brazen tablets" from "taking hold," as so many letter writers wrongly predicted it would. On a simpler level, the recurrent idea of the phonograph-clock continued to mediate between the largely discarded utilitarian purpose of the phonograph and a sense of having time for leisure. The musical phonograph helped define leisure time and space: in 1907 Edison's National Phonograph Company netted more than a million dollars on phonographs and prerecorded musical records for home amusement, even after skimming another million off to pay Edison for use of his patents.

Many idea letters proposed hybrids, like the phonograph-clock, and suggested merging communicative or inscriptive technology in ways that reveal collective attentions to issues of experience and materiality like the attentions so forcibly evidenced in the world of shorthand reporting and dictaphone sales. After the combined phonograph and motion picture machine noted above, the most frequent proposal was probably the typewriter-phonograph. "It occurred to me the other day," wrote one man, "that it might be possible to combine a Phonograph & a Typewriter so that a man (or woman) should only have to speak into the instrument & the words should be printed as uttered" (Hammaford 1890). For decades this device, the "typeophone" another writer dubbed it, seemed possible, new, even likely. Writers envisioned a dictaphone attached to a Remington or an Underwood, with no secretary in between, a purely mechanical conduit from the human mouth to the printed page. One writer, a budding reformer, went so far as to offer "a perfectly phonetically spelled language," which she or he felt the machine would require (Donaldson 1915). It was to be a purely mechanical conduit, so it needed a more natural language. According to others, the typewriter might be similarly joined with a telegraph or a telephone. One man inquired pointedly, "If by the agency of electricity a machine clicking or a word spoken at one end of a wire will be reproduced at the other end, why would not the same force hold good with a *typewriter*?" (Heath 1888). What these letters demonstrate, beyond the simplicity of their authors' grasp on the mechanics of phonographs and typewriters, is a willingness to unify oral and inscriptive action and a desire to produce legibility from orality. One bit of machinery assumedly worked just like the other, a resemblance that many thought must mirror a functional likeness connecting spoken and written words. Each potential typeophone embodied a theory of language only somewhat less reflective than those of Isaac Pitman, Melville Bell, and Alexander Ellis.

Similar combinations were studied and proposed for the phonograph and the telephone (an answering machine), the telegraph and the camera (a facsimile machine), the kinetoscope and the phonograph (talking pictures). Individuals proposed phonographs that could read from the pages of books written in braille, telephones that could display pictures of the person speaking at the other end of the line, and a multiplicity of other hybrids. Some suggestions were ridiculous, others perspicacious; some impossible, others already accomplished. Whether reasonable or chimerical in hindsight, ideas like these raise provocative questions about technological potentiality and any single, emerging sense of representational practice.

How and why did impossibilities like the phonograph-typewriter possess and retain their aura of possibility? How did practicalities like talking pictures or answering machines possess novelty within the folk culture of technological potential when they were already publicly receiving attention from inventors like Edison as well as from the popular press? Answers to both questions attest at least to contemporary experiences of inscription, representation, and reproduction as unsettled and mutually defining categories, multifaceted processes, decisively, if problematically, aural and visual. They further confirm that the folklore of technological potentiality is a long-lived affair, retaining its interests with the tenacity of those stargazers who so long supposed they were watching the Edison Star.

Multiple innovations to the means of producing and transmitting images and texts, whether accomplished or projected, would inevitably suggest similarities among practices of representation and means of perception. Orality, aurality, and visuality would huddle together under one potential umbrella of marvelous future technology. New inventions were necessarily understood in terms of old. Thus was the phonograph oddly considered a "witness" even though it is not an optical device. And when Edison filed his first motion picture patent caveat (a formal expression of intention to invent) in October 1888, he claimed an invention that "does for the eye what the phonograph does for the ear."[21] His "kinetoscope" would locate the common denominator of vision and hearing, just as the phonograph had located the common denominator of sound and text. Similarly, the accomplishments of 1876–1878 had been described this way: "The telephone is for the ear what the telescope is for the eye, the phonograph is for sound what the photograph is for light, and the microphone finds its analogue in the microscope," making minute stimuli perceptible to the human senses.[22] These and similar representations of new technology rely upon assumptions about the human body and mind, readily documented by an interesting subgroup of idea letters. Some writers explicitly concocted mental models. In 1896, for instance, a Mr. Fitzpatrick wanted Edison's opinion after reading:

A short time ago, of an English scientist who was experimenting in the new Xray. The account stated that this scientist took a picture of his own brain while thinking of a little child who was dead. When he developed the plate he found that there was a faint impression of the child of whom he was thinking when he took the picture.

Another man wrote in 1891, having read about but never seen the phonograph, that it had occurred to him "that the convolutions of the human brain are largely though not entirely a Phonograph," while the "grey matter acts . . . after the manner of the wax cylinder." He suggests further investigations, supposing that "a reading" could be made of the brain after death, useful in the case of murder victims and "sensational" in the case of an Egyptian mummy (Ivatts). For both men technological possibility is a function of media-memory; their points of departure are things that they read "a short time ago."[23] Both grope toward mechanist models of cognition for which new technology provides a metaphor, want thinking to be a form of representation and reproduction, and accordingly think anatomy is inscriptive. To these letter writers must be added the related case of the phonograph-afflicted, individuals who heard voices and, attributing them to the phonograph, wrote to Edison to be cured. Equally curious are the individuals who feared their privacy was being invaded by phonographic or telephonic devices. Like the "unsystematized" Ike Isacson in Elgin, Illinois, these individuals wrote letters that demonstrate the unreckoned power of technology as a metaphor within a culture (or an asylum) that internalizes surveillance as part of its discursive operations.[24] One man from Liverpool complained that he had "been Plagued this last 5 Years by an Notorious Gang of Murderers In the Shape of Men With a Phonographic Mind reading Machine" (Coyne 1888). Another wrote from Iowa that he had "discovered that an air telephone is in use, . . . Lots of times I get good information from it," he assured Edison, "but it gives me a headache" (Berg 1888). Though never in large numbers, evidently sincere letters like these made their way to West Orange every year. They are in no way representative, yet they serve as reminders of the stakes that rapprochements between technology and representation always involved—emotional, cognitive, semiotic, and more. Just as the camera obscura and then the steriopticon provided powerful metaphors for the status of the observer in the early and mid-nineteenth century,[25] the X ray, the phonograph, and similar inscriptive devices provided fertile ground for and cultivated emergent understandings of representation toward the end of the century. In comparing the brain to the phonograph, or even mistakenly identifying a phonograph as the origin of individually possessed pathologies, letter writers grappled with what it was to think, exhibiting a preoccupation with the sort of thinking that makes recorded voices and inscribed images intelligible as experiences and as evidence. Of course the vast majority of letter writers appealed not to cognitive models but to commerce. In doing so, however, they too revealed the unacknowledged weight of inscriptive technology as a source of orientation.

One pertinent idea that seems to have been suggested with greater frequency than either the phonograph-clock or the typewriter-phonograph was a sign system or annunciator for streetcars. Individuals wrote suggesting either electrically displayed signs or phonographically prepared announcements that would indicate when a trolley had reached individual stops or major cross streets along its route. The frequency with which automatic streetcar signs and annunciators were proposed as both new and desirable proves them to be reservoirs of cultural meaning in some measure characteristic of the late nineteenth and early twentieth centuries, when within half a generation, trolleys displaced horse-drawn cars.[26] In the very least it suggests that individuals feared—for themselves or for others—that the electrified streetcar would whisk them past their destinations unannounced and without their knowledge. They worried that the very vehicle by which they reached desired locations could be an agent of dislocation, that their best efforts at orientation would end in disorientation. In part their fear was a fear of cities, of the swirling urban chaos that Frank Norris characterized as one big "midway carnival" in his novel *The Octopus*. But streetcars existed beyond the cities as well, and streetcar signs and annunciators were more than an expression of urban fears and pleasures. As another novelist proved for a magazine article, you could get from New York to Boston by trolley if you had enough time, money, and patience for transfers. Theodore Dreiser made the trip in 1899 and reported that it could be done in thirty hours, with fifty-four payments of a nickel fare, and only a handful of gaps that had to be traversed by railroad or by foot. Dreiser maintains that at most points along the way, "No one seemed to know that there was any trolley connections anywhere save between their town and the nearest city" (Hakutani, 2:99).[27] Streetcars were in this way a powerful manifestation of the integrated local environment, articulating connectedness across the urban/suburban/rural span as effectively as the railroad had expressed connectedness across far more sweeping geographical distance. The improvements suggested for streetcars necessarily concern experiences and inscriptions of local integration and orientation. They turn out to offer an inverted imagination of the public sphere, one dependent as much upon the circulation of people amid print as upon the circulation of print amid people.

One letter Edison received, for example, proposes "a Street directory for Registering the Streets, to be installed in Street Cars which does away with the conductor calling off streets which means a Great benefit to the Public" (Anderson 1915). Though A. E. Anderson does not specify either electric signs or annunciators, the words "directory" and "registering" show that Anderson had in mind the sort of information control provided by city directories and mechanical registers of various sorts. "To register" streets was a verb that other letter writers also used to propose the same idea. Like those who wrote proposing the possibilities of a phonograph–cash register, a weather recorder, gas and electric meters, or a "register" to measure the number of people entering or leaving a building, Anderson was concerned with processing information mechanically. The benefits of handling such data mechanically were self-evident to another advocate of streetcar annunciators, who confided in Edison, "You know how many blunders are made and the annoyance and inconvenience by the present system. Breakmen [sic] speak so indistinctly"

(Minor 1907). A. E. Anderson wrote in 1915 from Deary, Idaho, a town that did not appear in the United States Census until 1920, and then with a population of only 316.[28] In her (or his) remote county of northern Idaho, Anderson may have been recalling her past experiences with streetcars. Or perhaps Anderson imagined the streetcar completely, savoring the technological potential of an entire vehicle at the same time that she advocated the possible "directory for registering streets." Whichever the case, Anderson's idea letter was an attempt to put Anderson on the map, to define herself. She sought acknowledgment, recognition, "a liberal offer," and in seeking these things proposed a device that literally located people, acknowledging their coordinates in space by pairing inscribed cross streets with trolley lines. Another writer suggests the extent to which space could be inscribed. Fred Strassweg's streetcar annunciator consisted of phonograph records installed "at all stations or streets to be registered." These records would be played when the trolley, with a "sound producing horn in the car" and a "sound producer" on its pole, arrived at the station or cross street (Strassweg 1907). In effect, the trolley acted as a phonograph needle and reproducer, playing a sequence of records on a jukebox that was as large or small as the entire streetcar route. The trolley would read its way along the streets, availing each passenger of the same paratactic text, pairing each available disembarkation point with a corresponding aural experience until such a time that the desired destination arrived in the company of an experience more satisfying than its predecessors.

The curiosity of Anderson's missive, proposing this "Great benefit to the Public" when the frontier village of Deary could not have contained the sort of "public" she (or he) had in mind, may inhere as much to Anderson's idea as to Anderson. The persistence of electric signs and automatic annunciators reveals a sense of dislocation, of disorientation, for which such devices offered a remedy both practical and metaphorical. Streetcar signs and announcements appealed metonymically; they stood for the limited geographical orientation of individual landmarks, yet they were attractive because they processed spatial information into discursive terms. And unlike the brakeman or conductor with his occasional discourse, they ensured that those terms were safely inscribed, predictable, distinct, and reproducible. Streetcar signs and annunciators provide one small indication of the extent to which modern space came to possess an acknowledged, new, public legibility. They propose a layman's language of space, apprehended by every streetcar passenger and imagined passenger. This language compliments the more arcane spatial discourse involved in the elaborate train signaling systems and devices required by the midcentury boom in railroads and suggested with relentless energy in letters to Edison, who made his own relevant contribution in the form of inventions like quadruplex telegraphy. Nor were signaling devices the most potent features of the modern discursive environment, which requires a whole "ethnography of reading." The phrase is Jonathan Boyarin's and stems from his admonition, "We will not get very far with a simple notion of 'reading' as a technology or practice independent of social context" (6). Like the railroaded American landscape and like technology (pace Boyarin), the practices of reading itself underwent significant changes during these years. In brief, more people did more reading as the years passed; they also read differently than they had before. Railroads and trolleys, for instance, provided modern, technological spaces in which to read. Newspaper and chapbook publishers invented editions specifically for rail passengers, and a new sect of advertiser monopolized the inside and the outside of streetcars with printed placards. Toward the end of the century, electric lighting dramatically increased the time and the place of reading, altering the context of oral and inscribed discourse (just as the fountain pen had changed the context of writing, making it an anywhere activity for the middle and upper classes). Reading was no longer a daylight activity. Electric illumination could even be defined in terms of legibility. As one Russian periodical boasted of an early illumination project of 1883, "In every point of Nevskii [Prospekt] it was possible to read easily." Urban space in particular became differently legible, due in part to street illumination and lighted store windows, but more so to the development of electric signage. A new "visual discourse," to employ David Nye's term, offered individuals a chance to read the city. While many features of this modern discursive environment could hardly have been visible from Deary, Idaho, letters from individuals like Anderson confirm the extent to which legible space was a widely shared experience, in part constitutive of a "public" and a "public benefit" stretching well beyond local population or experience.[29]

~

Some of the ideas Edison listed in his "Private Idea Book" or scrawled into his science fiction notes are the very same ideas that would-be inventors wrote to him again and again. Yet technological potentiality is differently constrained within the various discourses of literature, letters, and the laboratory. With many of the same technological preconceptions, Edison and the letter writers wrote under complementary misapprehensions: Edison seems to have thought that writing a novel would be easy, something he could do on the side. The letter writers supposed that invention was a possible sidelight, a self-expressive and renumerative act readily available to isolated individuals. Edison's literary attempt failed because the letter writers were wrong. However it might have been represented in the press, invention was Edison's full-time job, one that seems to have relied upon the collected expertise and institutional features of the laboratory he built, even as much as it depended upon any "private idea." Though misguided in their understanding of invention and occasionally eccentric in the inventions they propose, the letter writers need not be dismissed wholesale as the spokesmen of technology in America. They are volunteers conscripted to an accidental genre. This genre constitutes an authentic, collective response to contemporaneous features of American culture as varied in scope and as diffuse in character as the emergent context and meaning of technological change, of developing practices of representation and information control, and of Edison's vaunted stature as "The Wizard of Menlo Park." They demonstrate that popular conceptions of technological change were offset by continuities in perceived power relations, in patterns of epistolary utterance, and in the dogged maintenance of certain technological possibilities as new, notable, and desirable within what amounts to a folk tradition of technological "wisdoms." This unconscious culture of letters reveals a word-of-mouth folklore marked by long-open questions and long-held desires. As representations of potential technology, the idea letters attest to the centrality and complexity of representation itself in the face of increasing mechanization. They further attest to the unacknowledged weight that new technologies of inscription possessed as different sources of metaphorical orientation and psychical balance in a climate of change.

In one sense, the pre- and misconceptions surrounding the phonograph and other inscriptive devices of the late nineteenth century all suggest the same thing, no matter who was doing the pre- or the misconceiving. The stenographers who felt threatened by the phonograph, the men who promoted it as a dictaphone, and the many thousands who must have dreamt privately and earnestly of phonograph-clocks, phonetic typeophones, and streetcar annunciators all took a page from the same source: a sense of apparent inevitability, of modern experience and discourse made more material, and of modern material made more technological. It was a feeling not unrelated to their orientation as public beings, as citizens who were called by clocks, beckoned by signals and signs, quoted verbatim and entered into records, whether of tinfoil, wax, or paper. All seemed to sense that action and experience needed mediation and regulation in ways that available literacy practices would have to be stretched to encompass. Each newly imagined language machine, whether envisioned by an expert or amateur, with sincerity or literary license, answered an unasked question and assumed an unspoken theory of language and climate of representation. Ideas like the typeophone and the electric streetcar sign asserted the value of materiality, of texts as evidence. Ideas like the phonograph-clock and the streetcar annunciator asserted the value of regulated action and public norms, even as imagining the phonograph-clock may also have meant relishing the subversion of those norms. The currency of such ideas, as ideas, purchased a sense of orientation that was accessory to and helped make sense of the growing and pervasive culture of mechanical amusement. In another sense, however, the letter writers were less collective. Their letters are powerful evidence that experiences of technology are never as defined or definable as the history of technology makes us think. Each writer was a consumer, a receptor for the notion that modern technology solves problems, yet none of them consumed technology uncritically. If they had been uncritical, then they would not have written. Isacson, Anderson, and the others may all have listened to the phonograph, but they were each hearing different things. They consumed, but not without producing their own meanings. "Users," as one French critic puts it, "make (*bricolent*) innumerable and infinitesimal transformations of and within the dominant cultural economy in order to adapt it to their own interests" (Certeau, xii). The idea letters each tinker at inventing as much as they tinker with the social meanings of invention. They are appropriations, idiosyncratic creations of new technology individually formed out of old technology, hazy memories, misreadings of news accounts, experiences, and emotions, all in the presence of catalyzing personal interests that are both preconscious and conscious. The letters are valuable reminders that those interests exist, even when their articulation is imperfect or retrospectively obscure, even when they did not coincide with the mainstream interests of their time or fall later within the teleological optic of history. Personal interests, including those of appropriation and resistance, remain part of the everyday experience of technology. Idea letters are hardly a thing of the past: in 1992 astronauts on the space shuttle *Endeavor* repeatedly failed in their attempts to secure an orbiting satellite. After their troubles had been reported in the press, NASA started to receive unsolicited suggestions from individuals, who reached the agency by fax. One writer invoked the exemplary persistence of Edison. Another concluded with the reminder, "Of course, if you use my idea and it saves you money, I certainly would like to be compensated." Then in 1996 a rockslide in Japan left a remote, coastal tunnel crushed, with twenty people trapped inside. During the next week television audiences watched rescuers as they tried to dynamite through a gigantic boulder on top of the tunnel without causing further damage. Thousands of people reportedly called their TV stations with suggestions for the engineers.[30] In both of these cases there was a clear coincidence of personal and mainstream interests—catching that satellite, saving those victims—but there must also have been other, more idiosyncratic interests at play. Faxes to NASA suggested solutions involving knotted bed sheets and chewing gum. Individually they possess a possible irony as elusive as Isacson's possible sanity. Each demonstrates a vital and particular interest in technological knowledge, as well an interest in its accessibility, in the credo "you don't have to be a rocket scientist," to be a rocket scientist. This turns out to be a politically and epistemologically radical position, given the investment possessed by our dominant cultural economy in defining and authorizing technological knowledge by its own logic. The next chapter examines the exercise of that authorizing logic.

3

Patent Instrument and Reading Machine

When the United States Patent Office received Edison's application for a patent on the first phonograph, on Christmas Eve, 1877, official examiners confronted a device Edison and his attorney called an "Improvement in Phonograph or Speaking Machine." "Improvement in" may seem to represent an uncharacteristic modesty on Edison's part, but the phrase is actually a reminder that all granted patents are as much about old news as they are about new. Patents arise and exist within a highly regularized textual system of official filings, government notifications, and legal process.[1] Their knowledge sources are doubly the inventor's creativity and the examination of what is called *prior art*, an historical inquiry made by inventors, lawyers, patent office employees, more lawyers, and possibly the courts. Knowledge about both what is claimed to be new and what is discovered to be old is processed together, and if everything works out, new technology is authored and authorized: the inventor receives the right to a piece of intellectual property with possible, negotiable worth. In the case of Edison's "Improvement in Phonograph," Edison got his patent in mid-February without much trouble.

Preconceptions about technology, so evident between the lines of idea letters, science fiction, or commercial promotion, form the explicit method of the patent office. Edison's application, in order to be examined, had to be assigned to a division and class within the government's system of technological knowledge. At the same time, it had to do what every successful application for a letters patent necessarily (i.e., by law) must demonstrate: novelty. Indeed, the inventor may be naming something that does not yet exist for the examiners, who will always need to consider it within a taxonomy of their preexistent and therefore essentially faulty divisions and classes. For want of a better spot, the patent office put Edison's phonograph into Division XVI, Class 73, Measuring Instruments. Undoubtedly, this designation was made in order to avail the government examiners of Sub-Class 34, Recorders, where the phonograph ended up. Though not always to such an obvious degree, every patent application tests the system of knowledge previously bureaucratically schematized. And every patent grant incrementally alters the hierarchical structure of knowledge by changing its contents. The phonograph shifted both the substance and the relative boundaries of Sub-Class 34. Eventually the patent office has to bend under the weight of accumulated minute adjustments, revising its divisions and classes and thereby renewing the grounds of their contrived obsolescence. Only in 1886 would the patent office create a new class called "Acoustics," intended to include separate subclasses for the likes of phonographs and ear-trumpets. In 1890 the subclass for phonographs was changed to Graphophones after agitation from Edison's competitors. This change reveals how the classification of new inventions is muddied by commercial politics while it negotiates new technological knowledge.

The government's designation of the phonograph as a scientific instrument for "recording" measurements possesses many interesting implications, some of which I will defer until my next chapter, which is more explicitly about names and labels. In the present chapter I continue to address issues of text and evidence, interrogating these alongside matters of amusement and commerce. By now it will be obvious that I see the period around the turn of the twentieth century as one of particular upheaval and importance in the relations between words and things. Books and other printed text-objects became differently produced and accommodated within the nascent mass culture. Their becoming had much to do with the economics of the publishing trades, shifting grounds of authorship, new technologies, and new marketing means. It also had much to do with a welter of new inscriptions that had to be con*textualized*: to be defined against, and mutually to define the printed word as well as its established oral, aural, and readerly relations. These new inscriptive forms were cultural productions, like musical phonograph records and silent films, but they also included nonaesthetic products such as X rays and mimeographs. Between the Patent Act of 1870 and the Copyright Act of 1909 new relations between text and technology were clumsily negotiated, in particular, in the exercise and defense of intellectual property rights. The American patent process was one groundwork from which the new understandings arose; it provided a baseline, a relatively stable set of assumptions about words and things that, though they remained unquestioned within the confines of the U.S. Patent Office, were increasingly taken to task by the culture at large. Copyright legislation proved a much less stable ground. Congressional hearings and court decisions questioned the nature of "reading" in an effort to rearticulate the definition of constitutionally protected "writings." Generating particular controversy was the issue of musical copyright. Debate centered around the issue of whether phonograph records and perforated piano rolls could be "read." The same question was raised at the Berne Convention meetings in Berlin and was litigated in Mexico and across Europe. Legal case history in America focused on the copyright for a "coon" song entitled "Little Cotton Dolly"; the phenomenon of the recorded coon song itself implies some of the complex features of ownership and authorial identity that emerged as pressing questions of cultural production during the period. American lobbyists and legislators addressed the changing requirements of intellectual property statute within the context of changing technology and tastes, but also within the politics of the newly global entertainment economy that was enabled in part by varying social practices of consuming representations of racial, ethnic, class, and national difference.

Though here addressed consecutively, one of my points is that patents and copyrights belong together. Not only did the framers of the Constitution conceive them together, "to promote the progress of science and the useful arts," but during the first two decades of the twentieth century these forms of intellectual property were metaphorically situated and judicially constructed in a manner that drew them more closely together than before. Their proximity was to a small degree accidental, due to certain judges sitting on certain courts and to broad analogies that surfaced in the legislative history of the new Copyright Act. More so, however, their proximity arose from the commercial circumstances surrounding consumer goods such as cameras, phonographs, pianolas, and projectors, all patentable products reliant upon patentable supplies (films, records, music rolls) that in turn were reliant upon cultural products, the reproduction of original creations (the photograph, the musical score). Such consumer goods coevolved with complimentary elements of corporate practice, economic organization, and other production values. These together provoked a reckoning within the public sphere, not only of the nature and extent of ownership and authorship but also of identity and perception, of readers and writers.

~

The word *patent* derives from the Latin, meaning disclosed or lying open. In the early English use, *letters patent* meant an open, public document granting land rights or similar privileges to an individual or corporate body for an explicit period of time. Notably, British entrepreneurs were granted royal letters patent to found colonial American settlements in the seventeenth century. Territoriality has always been an operative metaphor in the rhetoric of patents, particularly in the articulation of the *claims* of a patent. Like a miner staking a claim, a patent applicant must demarcate the boundaries of her or his invention, seizing a metaphorical topography within a field of knowledge and an area of expertise. Edison made the point emphatically, when he instructed his attorneys to file a patent application for phonograph records; he directed them to "Claim the Solar System." In another similar instance he advised haste: "Claim the Earth, &

before she makes many more revolutions." Territorial metaphors were a point of legal comparison too. As one standard treatise of the early twentieth century put it, the patent claim "may be likened to the description in a deed which marks the bounds of a parcel of land"; though because the claim "deals not with a tangible thing," it remains "merely an approximation, more or less remote, to an exact disclosure of what a certain mind has accomplished."[2] Though the estimable Macomber here admits that "exact disclosure" is impossible, the rhetoric of patents rests squarely on the opposite assumption. The patent process assumes a naive relation between words and things. According to law, any "useful art, manufacture, engine, machine, or device" may be protected by "describing the said invention or discovery, clearly, truly, and fully," and undergoing a successful examination by the patent office.[3] Any technology can be described and any invention can be neatly, if not uniquely, divided into individual claims, that is, bite-sized acts of objective signification for which language is presumed to be clear, clean, natural, and free from ambiguity. No symbolic or figurative meanings pertain; the whole is exactly equal to the sum of its parts. In an age that imagined so many machines for language, the patent process offered a language for machines.

Edison's number 200,521 is a happy example of a patent to have at hand. While the patent office at first conceived the phonograph as a scientific instrument for recording; the genre of the patent document, which the patent office maintains so assiduously in its rules of procedure, was conceived as the textual equivalent of just such an instrument. Text-as-instrument and genre-as-technology are nothing new. Legal process requires written instruments. Shorthand, subsumed within its own technocratic pretensions, tried to be as much of a technology as possible. And many literary and nonliterary genres have been profitably considered in similar terms. Steven Shapin, for instance, has characterized the early printed works of the British Royal Society as possessing a "literary technology of virtual witnessing," showing that the genre of the scientific article has its roots in the need to attest, whether directly or obliquely, to the

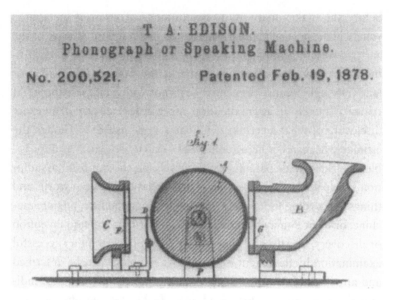

FIGURE 5. New knowledge. Edison's first phonograph patent, no. 200,521 (1878).

veracity of experienced events (Shapin, "Pump and Circumstance," 490). The article is technological in the sense that it is constructed, like a good scientific instrument, to lend weight to the findings it offers. Though the patent may have less to do with witnessing than the scientific article, the two share a similar intellectual history and are the authoritative discourses of a technical knowledge.[4] Like the authority of science, the authority of patents relies on objectively describing what really exists. But as harbingers of commercial advantages of one sort or another, patents fester with potential threats to objectivity. This is the "paradox of patenting." The patent system is supposed to stimulate inventive activity by inhibiting the diffusion of inventions.[5] American technologists have long recognized the paradox. A few, like Benjamin Franklin, declined to patent; instead, he published a description of his "Franklin" stove and encouraged its diffusion. However, as Thomas Fessenden explained in 1808, an inventor is usually "induced to divulge no more of the process than what might be deemed absolutely necessary to entitle himself to a patent; and confine the principles of his invention to his own workshop" (x). What this means for the patent document is a double-edged sword of specificity and vagueness. Inventors must "open," or disclose, their ideas so that no one can steal them, and in so doing, must describe their ideas minutely. Yet, they must also withhold as much as they can get away with and leave themselves covered in case of many possible unforeseen alterations. This is a pair of contradictions that may seem more in keeping with symbolist fiction and romance than with any technical discourse. The patent means to keep secret the very thing it means to reveal.

The paradoxical nature of patent documents arises separately from two connected points, the anatomy of the documents themselves and the activity of readership they imply. Anatomically the patent is a formulaic combination of three primary parts: a specification, a drawing, and claims. These primary parts are in addition to a colophon or preamble identifying the applicant and her or his nationality, as well as the patent number, date, title, and signatures of the applicant, attorney, and witnesses—features no less important to the document, but transparent in their rhetoric of pro forma validation. (Patent models were not required after 1880.) Each of the three primary parts requires consideration independently, as each possesses a coexistent and potentially contradictory role in the rhetoric of the whole. That whole is tempered by a jumble of implied readers similarly coexistent and similarly at odds.

The term *specification* occasionally refers to the entire patent document, but usually it has the more narrow meaning of a detailed description of the invention that accompanies a patent drawing and precedes its claims. Statute requires that the specification be prepared "in such full, clear, concise, and exact terms as to enable any person skilled in the art or science to which it appertains, or with which it is most nearly connected, to make, construct, compound and use" the invention. Macomber instructs his readers that an ideal specification should not waver from this purpose and "should avoid all laudation and all attempts to declare the basic character of the invention" (68). This ideal, Macomber admits, is freely avoided in practice. Edison's first phonograph is specified in his patent by three distinct gestures—an introduction citing the purpose and general nature of the invention; an explanation of its operation, with reference to the accompanying Figures 1 and 2; and an amplification of further means to the same end, with passing reference to Figures 3 and 4. Within his introduction, Edison even makes himself the subject of one paragraph, narrating that "after a long series of experiments" he has discovered that the human voice produces "separate and distinct" vibrations in a diaphragm "or other body," and "therefore it becomes possible to record and reproduce the sounds of the human voice." Such an explicit narrative, an almost autobiographical digression, is unusual and unnecessary in patent documents; most patents do not look like the stories of inventions or inventors. It is this supposed lack of narrative logic that likely accounts for their reputed dullness. Patents are pretty dry reading. In fact patents *do* demonstrate an aggressively narrative logic; they all imply Whiggish accounts of technology, the gaps and misperceptions of the past cleverly remedied by the invention at hand, which will be of certain use in the future.[6] This underlying narrative is a rhetorical prerequisite for any patent, a carefully couched assumption that emerges in the patent specification and claims and to which, in granting any patent, the United States government accedes.

The specification of Edison's first phonograph patent dilates on some possible alternatives with the phrase, "It is obvious that many forms of mechanism may be used." Edison describes several alternatives, but indicates that much more remains undescribed. They are already manifest, self-evident. The rhetoric of self-evidence is implicit in all specifications, which frequently revolve around similar statements. This rhetoric articulates the patent's implied reader, "skilled in the art," who fully understands the invention and its implications from the specification given. The implied reader is a qualified reader, equally expert in reading patent documents and in the "art" at hand. Even as it posits this qualified reader, the rhetoric of self-evidence excuses the limits of the text. The author and the implied reader collude in their acceptance of alternatives and specifics that, because "obvious," remain unwritten or unexplained. Particularly after the Patent Act of 1870, when photo-lithographed drawings became a standard accouterment to patent documents, the patent specification typically functions as an elaborate caption to the drawing, indicating features of the invention by reference to letters or numerals on the drawing itself. Like the specification, the drawing is addressed to persons "skilled in the art." It need not be a working drawing but participates, like the specification, in a rhetoric of self-evidence. As a representational device, the patent drawing generally relies on mapping structural features of the represented object and is thus among the simplest of such devices.[7] Edison's drawing does not, for instance, seek to represent the vibrations of the human voice acting upon the phonograph diaphragm, but rather satisfies itself with an orthographic projection of the placement of the diaphragm within one version of the necessary machinery for recording and reproducing. Drawings are not always so modest—Emile Berliner's gramophone patent (no. 372,786, issued 1887) also shows no sound vibrations, but it does show an exaggerated lateral groove in the recording surface (the innovation of the gramophone), so that the features of his illustration map structural components of the invention in multiple scales. In figures of various magnifications the gramophone is disassembled, and in all cases where the lateral groove is shown, it is greatly exaggerated beyond the scale of the figure.

In selecting structural features for display and in disassembling and/or exaggerating them for the eye of the observer, patent drawings exist within a historical context of technical illustration. Sixteenth-century authors such as Agostino Ramelli and Georg Agricola explored perspective and developed "exploded" drawings as a means of communicating artisanal information. In seventeenth-century Britain, William Petty and other members of the Royal Society proposed a "history of trades," for which illustration became a necessary component, because, as Petty wrote, "bare words being not sufficient, all instruments and tools must be pictured, and colours added, when the descriptions cannot be made intelligible with-

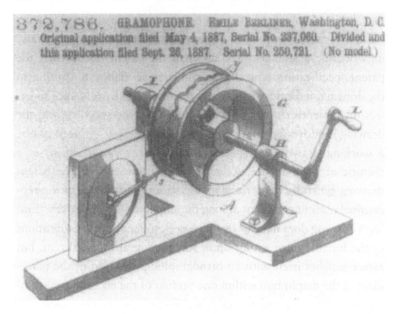

FIGURE 6. Emile Berliner's gramophone patent, no. 372,786 (1887). Gramophones differed from phonographs and graphophones in that sound waves were incised in lateral, side-to-side, grooves, rather than vertical, hill-and-dale grooves. These were two distinct operating systems with incompatible file formats.

out them" (qtd. in Ferguson, "Mind's Eye," 830).[8] Within this context, drawings do what language cannot. Today many engineers and historians of technology would concur that technological knowledge cannot be "reduced" to words. Technology possesses an essential component of "visual thought" and remains remote from language at critical moments of invention and design. Besides sharing the pejorative sense of shorthand authors' "common" writing, "reduction" into words joins with what the patent system calls *reduction into practice* in elevating the newly envisioned idea itself, rather than its written description or pictoral representation. Yet the American patent system contains no admission of linguistic insufficiency; drawings are complimentary to the specification's language. When the required submission of patent models was discontinued in 1880, it was as if to say that language suffices, clarified by and mutually clarifying the patent illustration. Like so much else about patents, this linguistic priority was articulated in the courtroom. For one, *Edison Electric Light Co. v. Novelty Incandescent Lamp Co.* tested the relative weight of patent drawings and specifications. An Edison lamp patent (no. 444,530, granted 1891, reissued as no. 12,393 in 1905) described and claimed a lead-in wire for electric lamps that contained segments of copper and platinum, joining them within the glass at the base of the bulb. His arrangement ensured strength and conserved platinum, a major expense in lamp manufacture. But a similar joint had been shown by mistake in the patent drawing for another light bulb (no. 401,444, granted 1889). The Lemp and Wightman patent did not specify the glass-encased joint, but their draftsman had mistakenly pictured it. In 1909 the courts upheld the validity of Edison's patent, indicating that the mistake was just that, a mistake. Anyone "skilled in the art" could see from their language that Lemp and Wightman had not invented or patented the invention their draftsman had pictured. A mistaken drawing would not invalidate a patent because, given a specification and its qualified reader, language is sufficient to represent technological knowledge.[9] Bureaucratic procedures and further litigation continued to articulate the relative weight of patent drawings.

If the patent document was merely the description of a machine or mechanical process, then the specification and drawing would be sufficient. Patents assert property rights, however, and as such they describe an innovation, an idea embodied in the machinery detailed by the specification and drawing. The patent claims enumerate this idea, dividing it into parts, marking its bounds, and employing a territorial rhetoric alluded to earlier. In effect, the claims present the inventor's (or her or his lawyer's) pointed *reading* of her or his specification, a reading checked and acquiesced to by the patent office during the application and examination process. As readings, patent claims explicitly identify the patentable features of the specified machinery, process, or design, frequently identifying them—like Edison's phonograph patent does—as "substantially as specified" or as herein "substantially set forth" in the specification. These and similar phrases, Albert H. Walker explains in his *Text-Book of the Patent Laws*, are "always implied in claims wherein [they are] not expressed" (171). Claims are always dependent upon specifications. Specifications offer descriptions necessarily cited by the claims, which generalize allusively from the particularity of specifications.

In keeping with their territorial rhetoric, the principal characteristic of patent claims is breadth. Claims are either "broad" or "narrow" in the parlance of patent seekers and the patent law. Because his "Improvement in phonograph or speaking machines" was the first invention of its kind, Edison was allowed four very broad claims. His first encompassed the most:

The method herein specified of reproducing the human voice or other sounds by causing the sound vibrations to be recorded, substantially as specified, and obtaining motion from that record, substantially as set forth, for the reproduction of the sound vibrations.

Most patents and patent applications hedge the inventor's bets by including multiple claims, usually giving them in order, like "a set of boxes each enclosed in the last, each getting more specific" (Myers, 75). Edison's second claim identified his invention:

The combination, with a diaphragm exposed to sound vibrations, of a moving surface of yielding material—such as metallic foil—upon which marks are made corresponding to the sound vibrations, and of a character adapted to use in the reproduction of the sound substantially as set forth.

Here Edison's invention is narrowed to protect his "moving surface of yielding material." He identifies such a material, but the example of metallic foil is not an exclusionary one. By comparison, claims can be exactingly narrow. In his "Improvement in Type-Writing Machines" (no. 133,841, issued 1872), for instance, Edison had been allowed eight claims, the second of which reads:

The rack-bar *h* and spacing-pins 7, in combination with the spring-pawl 8, key *d*, and type-wheel *a*, substantially as set forth.

This claim, like the patent specification, makes reference to the patent drawing. Though less clearly a reading of the patent specification, it does interpret the described machinery, identifying a specific feature that Edison and his attorney believed possessed the three characteristics required for a patent—invention, novelty, and utility. Linguistically, the narrowness of a claim inheres in its specificity, or rather in the reciprocal specificity of the specification it interprets. The dependence of claims upon specifications might suggest they are more important before the law, but this is not the case. If claims offer a reading of the specification, then patent law relies upon a metareading in which attorneys and judges concern themselves with analyzing the way claims interpret specifications. For instance, in his kinetographic camera patent of 1897, Edison specified a new and valuable invention, a combination of elements that constituted the first practical motion picture camera. Yet *Edison v. American Mutoscope and Biograph Co.*, known as "the First Mutoscope Case," rendered three of Edison's six claims void because they were too broad. The claims had been miswritten, the specification misinterpreted. Had he claimed the organization of mechanical parts represented in his specification, Edison would have been covered; instead he claimed "an apparatus" that did what his camera did, without noting the specifics. (Edison's first phonograph patent, which made similarly broad functional claims, was read with greater largess because it constituted such a deviation from the prior art.) Fortunately for him, Edison was granted a reissue of the patent, narrowing its claims. Even these claims were contested in the courts (the "Second Mutoscope Case") and another reissue was obtained. Ultimately, the matter remained in litigation until 1914. Throughout, Edison's camera was not in question; he had specified a very valuable invention. That is what all the fuss was about. The litigation concerned only his claims and the reading of his specification, which he continued to revise in reissued patents.[10] A sustainable patent necessarily offers a good reading of itself; a valuable patent can offer an excellent reading of itself. A patent's self-interpretive value and an inventor's rhetorical acumen may both appear directly proportional to the value of the patent generally, though plenty of worthless inventions have been specified and claimed with exceptional linguistic dexterity.

In focusing so specifically on patent claims, the law resembles the many correspondents who sent "idea letters" to Edison, writing as if ideas were tangible objects to be bought and sold, transparent in their extent and implications, the way a coin has shape and worth, or the way a machine has design and function. It is even possible to patent a technology that has not yet been constructed. Edison described and received a fundamental patent on a coiled carbon filament for his electric lamp before he had figured out how physically to fashion the filament itself.[11] If the technology has not been built and the patent "creates nothing" on its own (Macomber, 1), then what exactly is the subject of Edison's patent? It is his *idea* of the coiled carbon filament mounted with lampblack putty. Edison's idea determines his invention in the sense that it is supposed to have existed prior to his possession or embodiment of it, prior to his identification of it against the backdrop of all the "prior art." Like an undiscovered mining property, the idea lay undifferentiated from its surroundings until claimed. Nothing about claiming involves constructing new knowledge: patent law avers that "the breadth or the narrowness of a claim as the case may be, does not depend upon any artificial rule of interpretation" (Walker, 102). Instead, by implication, inventions themselves have the inherent quality of breadth or narrowness. They exist within a natural landscape of ideas, some discerned, others yet to be discovered, each taking up its own essential and appropriate width along the frontier. This is an extreme of what historians of technology call *internalist* reasoning; the extent of an idea's patentable "invention, novelty, and utility" is a natural (as opposed to Walker's "artificial") quality, inhering to the invention itself in its relation to the "prior art," rather than a construct, meted out by contemporary and socially determined parameters of novelty and usefulness, or, more specifically, by any discursive activity of the patent document and its readers or by any beneficence of the government, which grants rights but does not make property.

Much exists to contradict this internalist logic. The implied acts of writing and reading upon which the genre depends do much to preclude the notion of a "natural" or essential breadth or narrowness to ideas, even as they preclude the notion of a natural, straightforward, unrhetorical discourse in which objects or ideas can be cleanly, clearly, and fully signified. The whole patent document takes the form of an oath, addressed "To all whom it may concern" and sworn to by the inventor, so that in a certain sense the patent does resemble a scientific article, making public new knowledge ascribed to an expert or experts. The inventor is not its sole author however. Rather the "author" figure of a patent document combines the inventor, her or his attorney, and the patent office that has examined the application and granted the patent. The "author" of the patent genre is thus a rhetorical figure quite distinct from the inventor whose idea gets patented. Authorship is scattered. Meanwhile, the implied reader of the patent genre is both "whom it may concern" and the "person skilled in the art," to whom the specification and drawing are necessarily directed. The patent's title implies its first reading by the commissioner of patents. The claims attest to the manner in which the patent document reads itself, propounding the significance of its own specification in light of the "prior art." Finally, the patent document implies an additional, authorized reader, to the extent that the federal courts are its readers of last resort, turned to for decisions in cases where qualified readers—inventors, attorneys, and their expert witnesses—disagree on matters of interpretation. Like its author figure, the implied reader of patents is polymorphic, nested within the transactional process of application and examination, publication, and what Carolyn C. Cooper calls "patent management," or the selling, licensing, litigating, and reissuing of patents, extending and manipulating property rights after the grant and publication of the original document.

Nor are writing and reading mutually exclusive activities when it comes to patent documents. In particular, the federal courts combine the functions of reading and authoring. As *authorized* readers, the courts can change the texts they read as they read them. This holds true for the functional parts of patent documents as well as for the meanings of individual words. The courts author patents by writing them into the discourse of case law, usually in deciding infringement suits that patent holders bring to protect their rights. For example, case law stipulates the relative weight of patent drawings and specifications in *Edison Electric Light v. Novelty Incandescent Lamp*. Case law identifies when vague claims are too vague and when broad claims are too broad. Court decisions enter a fabric of citations and an ongoing negotiation of meaning. The First Mutoscope Case was cited in twenty-four additional court decisions as well as in legal treatises like Macomber's and textbooks like Walker's. Judge Wallace's decision erased three claims from Edison's patent and also indicated for all patents a breadth that was too broad; such limitation would continue to be tested and negotiated in other court rooms and in consideration of other patents. This form of authorized reading is clearest in the definition of individual words. The meaning of individual words within a patent document can be specifically identified by the author, or else devolves upon the imagined interpretive powers of a qualified reader. About these imagined powers differences frequently occur; the court then renders a decision that cements the word in question to a particular definition. The patent for Edison's first phonograph, for instance, turned out to be of surprisingly little value given the breadth of its claims, all because it repeatedly uses the word "indentation" to refer to the grooves made in the recording surface made of foil or other "yielding material." Ensuing legal process determined that later, wax-based recording surfaces were more properly "engraved" than "indented," since wax was actually removed and didn't just yield. Edison's patent lost its value as the technology and attendant discourse changed.

Reading patent documents is a time-based experience. Patent rights extend only seventeen years; after that, the meaning and significance of the patent document change. Though many genres exist in the climate of their own obsolescence (scientific articles, guide books, and so forth), few genres short of almanacs and messianic, prophetic texts have such a sharply defined metamorphosis built in. After seventeen years a patent's claims no longer mark proprietary bounds of new technological knowledge. Instead, the patent document as a whole attests only to old technological knowledge. It loses its performative quality as law and remains only authoritative evidence of the "prior art," turned to, if at all, during the birth pangs and adolescent contests of other claims in other patent documents, or turned to retrospectively by historians of technological knowledge. The former provenance of its qualified reader becomes more that of an interested reader. While the document continues its place with other documentary evidence, it is no longer a viable legal instrument.

The patent document is a remarkably formulaic and stable genre built upon active contradictions. For two hundred years American patents have declared in their rhetoric, "I am not rhetorical," "I tell no story," and "I am an expert at representing what exists," yet they have done this within bureaucratic and litigious wrangling that attests greatly to the contrary. Their confidence in "what exists" remains troubled by unreflective definitions of expertise and representation. Despite its pretensions to render better and better measures of an inventor's idea—the way that science is supposed to offer increasingly accurate descriptions of nature—the process of patent law reveals patent documents to be rhetorical instruments within a vast system of meaning creation, a "discourse" in the active sense, operating from the first draft of a patent specification to the decision of the ultimate court of appeals.

If according to its own rhetoric, the patent document offers a clear, clean representation of a new invention embodied in the technology it clearly and cleanly represents, then it is no surprise that patent documents generally take no special account of technologies which produce representations. Nor do the patents for representational products take any special account of what representation consists of, since the whole work of representing is assumed to be transparent and objective. Three patent drawings illustrate the limitations of this assumption: Emile Berliner's gramophone patent pictures an exaggerated lateral groove, not sound waves or the human voice. Edison's patent on kinetoscopic film (reissue no. 12,038, granted 1902) pictures a tiny strip of film that purports to show five "uniform sharply-defined photographs of successive positions of an object in motion," a gymnast in this case, though the "photographs" pictured are so close that they do not convincingly represent "successive positions" of the tumbler. Finally, Frank Lambert's typewriter patent pictures his machine typing out the words "I enclose check for Fifteen Dollars," the necessary patent application fee.[12] All three illustrations are of inscriptive products, and in the case of phonograph records and films (not typescripts), patentable ones. Each of the three in its own way represents an act of inscription made variously mechanical and does so with the differing degrees of literalism necessitated by the experienced aesthetic distinctions between forms. The printed page cannot represent sound or motion with the same literal facility that it presents typescript through the device of quotation. Nor can the patent document. However, the genre's (and the government's) confidence in its own legibility, which is to say, in the sufficiency of language, tends to deflect attention from the varieties and complexities of representational acts and products.

Notwithstanding this generic avoidance, the character and status of representational products were legally contested in a number of ways during the early twentieth century. Two of the most hotly contested points were addressed as matters of intellectual property. Tying and price-fixing were both strategies used by manufacturers to control the market and could be defended as natural extensions of the patent holder's right to "make, sell, or let for hire" the patented article. Price-fixing could additionally be defended as a natural extension of copyright. *Tying* referred to the practice of requiring distributors and consumers of a patented article to use only the patent holder's subsidiary products and supplies. Razor blades were tied to razors; paper and ink to mimeographs; film to cameras; records to phonographs. In such cases the razors, mimeographs, cameras, and phonographs were distributed and retailed under explicit license agreements or bearing explicit notices that required the purchaser to operate them only with that same manufacturer's supplies. Similar license agreements stipulated the minimum prices to be charged for copyrighted books or patented mechanisms and supplies, a guarantee that was aimed at stabilizing distribution networks (and

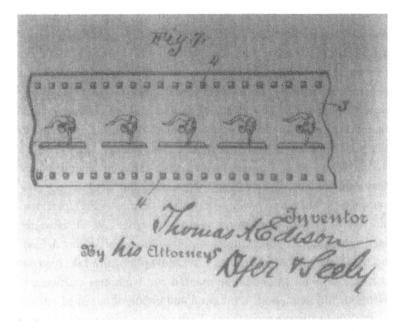

FIGURE 7. Motionless drawing of a moving picture. Edison's film
patent, reissued as no. 12,038 (1902).

profits) by forbidding intrabrand price competition. Anyone violat-
ing these license agreements was held to be infringing the copyright
or the patent. What this meant, in effect, was an additional topography adding more commercial dimension to intellectual property
that was already, in the case of patents, spatially conceived. Not
only did inventions possess breadth or narrowness but also they extended into the marketplace to greater or lesser degrees. When it
came to pricing and the supply of subsidiary products, an inventor's
rights might determine the actions of wholesalers, retailers, and
even individual consumers.

To be sure, different kinds of products and different sorts of commercial arrangements had always possessed the same implied topography. Network technologies, for instance, offer opportunities for
the inventor or the patent-holding manufacturer and capitalist to extend themselves into the market, while new forms of networks press

the same questions in different ways. Systems, whether of the telephone, telegraph, current distribution, or modem and Web browser, literally articulate the connections between producers and consumers that tying and price-fixing were aimed at affirming. Mail order and subscription sales of books and other items do the same thing, though these are notably forms of commercial organization rather than a feature of intellectual property or of network technology. In the nineteenth century, publishing houses such as Mark Twain's doomed Webster & Co. sent canvassers to different regions, where they exhibited a prospectus and tried to sell a book before its publication, as if they were selling shares in a company.[13] Department stores and "trade" bookstores sold the novels of W. D. Howells, Henry James, and others, but took up goods like Twain's *A Connecticut Yankee in King Arthur's Court* only piratically. Like mail order, subscription sales accommodated rural markets particularly well, but suffered a poor reputation and greatly diminished returns as the nineteenth century closed. They remained attractive as a way of controlling distribution and pricing, bettered by tying and price-fixing where intellectual property rights could be used as protection.

The legality of both tying and price-fixing was repeatedly tested. Neither issue had exclusively to do with representational products, yet many of the test cases and precedents involved copyrighted books or the patents for inscriptive devices, their subsidiary products and supplies. Quite a number of the price-fixing cases had as well to do with another new feature of the market, department stores, which aggressively used prices to attract customers and seemed to have early on realized the value of cultural forms like books and records as loss leaders, enticing customers through their doors. The department stores were perceptive; they reoriented retail with regard to price discounts and recognized the ascendancy of amusements in the emerging alignment of leisure and consumption. Mark Twain had fulminated against John Wanamaker's in Philadelphia for selling his books at a discount. Edison's National Phonograph Company sued both The Fair in Chicago and Kaufman's in Pittsburgh for cutting prices. Jobbers and dealers were made to sign elaborate price-maintenance agreements. Columbia Phonograph took on Gimbles and the Victor Talking Machine Company took on R. H. Macy's in New York. As it turned out, Macy's became embroiled in three important price-fixing cases, two that challenged the extension of copyright to cover pricing and one that challenged the extension of patent rights for the same purpose.

In 1908 and then again in 1913, the Supreme Court held that an author's copyright did not permit her or him to specify the resale prices of a book. Using these copyright cases as authority, the Court would later hold that similar prohibitions should extend to patent holders.[14] The judiciary proved more mercurial on the issue of tying. In 1909 the Supreme Court affirmed the legality of tying unpatented phonograph records to patented phonographs. In *Leeds and Caitlin v. Victor Talking Machine*, the Court decided that Victor could block the sale of non-Victor records for use on its Victor machines. The decision was made on the grounds that, even though an unpatented element in the combination of phonograph and disc, the disc was an active and durable partner in the combination, each playing of a record recreating, in effect, Victor's patented invention. In deciding thus, the Court felt it had to refer to a nearly parallel case that had been decided differently. In *Morgan Envelope v. Albany Paper*, paper supplies, unlike records, had been deemed passive and perishable or transitory. Similar distinctions between wax and paper as mediums of inscription were at issue in the legislative tangles taking place over musical copyright and must be addressed in the next section of this chapter.[15] *Leeds and Caitlin v. Victor* is an important point of contact between patents and copyrights, and its timing coincided with the new Copyright Act in 1909.[16] The decision effectively allowed the Victor Company to block the duplication of its records by the Leeds and Caitlin Company, which quickly went out of business. This was a long way from granting copyrights for recorded sound, but it did draw the recording studio under the aegis of intellectual property law by implying a distinction between duplicating and recording.

Firm precedents establishing the illegality of tying and price-fixing were not handed down until 1917, when the Supreme Court issued decisions in two important cases on the same day: *Motion Picture Patents Company v. Universal Film Manufacturing Company*, and *Straus [Macy's] v. Victor Talking Machine*.[17] These decisions have stood, though niceties of the same issues continue in litigation today, in suits about book distribution and new technological and corporate developments, like "bundling" computer software and operating systems with computers. For much of the period in question, however, products, including representational ones, drew a large measure of quality as commodities and a substantial measure of identity from the technological means of their production and use, from the "interdependence of use-values" escalating within the culture and economy.[18] Tying and price-fixing implicated cultural productions in the sale and use of their means of production. On the mechanical level this meant that films and records were defined by the conditions of their sale and operation with and by patented and explicitly licensed projectors and phonographs. On a corporate level the same interrelation of product and producer was routinely reflected in management structures and was mirrored in the complex economy of the nascent "star system," which offered the celebrated producer as a secondary product, as well as in the distribution networks established between manufacturers and consumers. Inscription, in this context, emerged as both emphatically mechanical and implicitly conflicted by the related pressures of antitrust (procompetition) sentiment and changes to the structures of wholesale and retail.

The genre of the patent document proved able to avoid all these complexities surrounding mechanical representation in a way that the judiciary and legislators could not. Western culture assumes that the scientific article and the patent document are instrumental genres par excellence. Both are counted upon to avoid ornament, to home in objectively on what exists, either by cleanly describing a true experience of nature or by cleanly describing the exact nature and extent of an invention. The instrumental status of both as genres remains curiously unquestioned by the ongoing revisionism of scientific practice, on the one hand, and the process of patent law, on the other, unquestioned because the rhetoric of both necessarily relies upon a simplified model of representation as clean, untroubled truth telling. But beneath an unassuming face, the rhetoric of each genre proves an intricate matter. Patents rely upon an implied, qualified reader, "skilled in the art," who unifies technological and textual expertise. Like scientific articles, patents identify new knowledge. They do so within a context that avers the knowledge itself to have been preexistent, waiting to be "embodied" in the "art," to be measured and described in a flexible, interpretive language game called *claiming*, in which the patent office plays a hand, federal judges act as umpires, and a seventeen-year clock ticks on the wall.

～

I wish to relate an incident which occurred in our store this evening, which I considered the highest compliment that has ever been paid to a talking machine. I was exhibiting a Home [phonograph], with a 24-inch silk horn. Among other Records, I put on No. 8656, The Flogging Scene from "Uncle Tom's Cabin." When the Record was ended, a man sprang out of his chair, wiped the tears from his eyes, and said, "I'd give ten dollars for the privilege of hitting that damned slave driver just once."

    —From the Buckeye Music Co., Rudolph, Ohio, in *Edison Phonograph Monthly*, February 1905

At the end of the nineteenth century the new technology of recorded sound helped to challenge the visual habits of musical practice. Audiences accustomed to watching performers, who might themselves be eyeing a printed score, could now hear music with nothing to look at but a piece of machinery. Phonographs, wax records, pianolas, and paper music rolls were all new commodities troubling the established musical trade, in part by questioning the visual norms of intellectual property. The phonograph record and the music roll had to be contextualized, to be located against the legible, copyrighted texts of lyrics and notation, which comprised the inviolable units of protected property under Article One, Section Eight of the Constitution. Legislative hearings and judicial decisions questioned the nature of reading in an effort to rearticulate the definition of protected "writings." Congressional debate centered around the issue of whether phonograph records and piano rolls could be "read," in what became an early and elaborate exploration of textuality in the new age of machine-readable text. While Congress tinkered amid the essentialism of American copyright law, honing its application to materially new cultural forms, the shifting optics of popular music brought pressure to bear on other visual habits, including associations between racial difference and skin color. By removing the performer from view, the technology of recorded sound also removed the most keenly felt representation of the performer's race. American musical culture engaged difference in new ways, provoked at once by the enormous popularity of racist coon songs during the late 1890s, by early attempts to delimit and commodify authenticity in so-called "Negro" music and by the phonograph itself, a mimetic machine that had not failed to accumulate its own parcel of racial associations in the several decades since its invention in 1877.

The above anecdote received from the Buckeye Music Company, which appeared in an Edison trade publication, provides some access to the less familiar elements of musical culture at the turn of the century. Music stores like Buckeye were distribution points for sheet music, song books, musical instruments and supplies, as well as phonographs and records. Phonographs occupied an ambiguous position as "self-playing" musical instruments. Without the benefit of radio broadcasting, potential customers had to hear phonographs and records in public in order to know and desire them, so that exhibitions and demonstrations like the one described were frequent and necessary events in showrooms everywhere. The man who jumped up and offered ten dollars to hit Simon Legree was paying the "highest compliment" to the phonograph because he offered to pay a ten dollar sum so in excess of the thirty-five cents that an Edison record of "The Flogging" really cost. Embedded in this narrative of paying compliments and paying ten dollars are a host of implications about mimesis, culture, and commercialism. The impassioned listener of "The Flogging" may or may not be mistaking fiction for reality. He does seem to know *Uncle Tom's Cabin*, whether from reading the original or from experiencing some of the plethora of adaptations that had appeared in print and on stage in the half century since Stowe's novel was published in 1852. Either way, the Buckeye proprietor feels complimented because his recording provokes such a powerful emotional response, the way reading the novel might, while the Home phonograph and its record cylinder remain relatively unattended components of the listener's experience. The man didn't hear the phonograph or the record; he heard *through* them to Simon Legree whipping Uncle Tom. It is this selective hearing that the Buckeye proprietor recognizes as the highest compliment that can be paid to any communicative or inscriptive medium, including the talking machine.

The proprietor's anecdote plays off of an important trope resident in Anglo-American constructions of race and class, the familiar narrative of the alien naif who mistakes mimetic representation for reality. In the fictions of Dickens and Twain it is the uncultured bumpkin who takes theatrical production literally. Many "true" anecdotes of this sort circulated during the nineteenth century. There was the Baltimore man who objected "to an assault on Coriolanus because 'three against one' was not a fair fight," and the man from New Orleans who suggested "to Othello, grieving over the loss of his handkerchief, 'Why don't you blow your nose with your fingers and let the play go on.'" The full truth of these anecdotes cannot be gauged, but the blurring of reality and mimetic action that the anecdotes relate is likely of equal consequence to the circulation and persistence of the anecdotes themselves. With each telling they present and assert culture as an exclusive activity for those who have it and "get it."[19] New technical cultures seem to have relied upon similar anecdotes in their construction of male, technocratic expertise. Nineteenth-century telegraph and engineering publications exhibit a rhetoric of exclusion on the bases of class, race, and gender. Their pages are filled with anecdotes about bumpkins who shimmied up telegraph poles to hear messages as they went by and about women who made other errors with regard to new communications technology.[20]

FIGURE 8. Type scene. The "Little African and the Too Versatile Phonograph." Edison received this cartoon from an admirer, who inscribed it, "A great 'ad' for Phonographs—eh?" The "African" destroys the phonograph only when it plays a coon song.

The same exclusionary trope extends into constructions of racial difference in exploration literature and ethnographic accounts, where racially distanced "natives" play the bumpkin's part, fooled by their own reflections in a mirror or—a type-scene in the same tradition—by voices emanating from a phonograph. From the 1880s through the 1920s versions of this type-scene filtered into travel narratives, trade publications, and newspapers, into cartoons, comic films, and serious documentaries. The politics of these interracial encounters was characterized by the self-congratulatory aggression of Western technological achievement and colonial dominance. They

seem geared to provide an accessible "comic relief" against which
Anglo-American culture could receive the less accessible accounts of
its resolute ethnographers, who bundled up phonographs and motion picture cameras for journeys to even less accessible climes, to
record the curious natives (in the double, us-and-them, sense of "curious"), all in the name of science. Phonographic and cinematic inscriptions fit the logic of ethnography exactly. The recording phonograph and the camera interceded between the ethnographer and his
subject, offering a rhetorically valuable sense of technological impartiality and receptivity. The resulting records and films concretized
what is now called the *ethnographic present* tense of anthropological description, freezing the ethnographic subject in time, providing
"live" recordings as specimens for further study at home.[21] If the
proprietor of the Buckeye Music Company only hinted in these directions, then the Edison and Victor phonograph companies did too,
making mimetic confusion a matter of kitsch in their respective promotional images of a well-dressed toddler breaking open a phonograph ("Looking for the Band") and the more famous dog listening
to one ("His Master's Voice"). These trademarks tone down and
make "cute" and commercial the exclusionary trope by substituting
children and pets for the aliens who where elsewhere fixtures of dis-
tancing between classes and races. As Michael Taussig explains, the
images succeed because they continue "to reinstall the mimetic faculty as mystery in the art of mechanical reproduction, reinvigorating
the primitivism implicit in technology's wildest dreams, therewith
creating a surfeit of mimetic power" (208). Edison's National Phonograph Company briefly offered its agents multiple electrotype versions of "Looking for the Band" to promulgate its mimetic surplus;
one was the familiar toddler and another was a pair of bug-eyed,
black-skinned caricatures.

I am suggesting that intimations of class and race politics lie embedded in the Buckeye proprietor's anecdote as published by the
Edison Phonograph Monthly. Furthermore, similar intimations of
question and conflict over matters of identity and cultural hierarchy
lay buried at different depths in the emergent culture of recorded

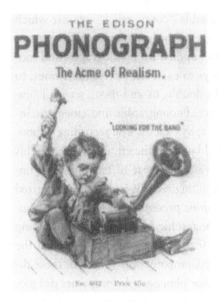

FIGURE 9. Promoting mechanical reproduction. Children, "natives," and pets help define the medium. *Above*, "Looking for the Band." Two different electrotypes offered to Edison agents for their own advertisements (1903). *Opposite*, "His Master's Voice" trademark (1900).

sound in America. This should be an unremarkable claim by now, the requisite extension of politics differently recognized in popular music by Adorno and other theorists. At the same time that the technology of recorded sound provoked a reconsideration of statutory authors and readers in debates over copyright, recorded sound helped to modulate the already Gordian politics of popular music. Recordings further complicated the identities of musical authors and performers. The new, hungry mimesis of the recording phonograph itself came to market larded with assumptions about sameness and difference, about cultural appropriation and assimilation. Consider the impact of the phonograph, which Dave Laing calls "a voice without a face," on the tradition of blackface minstrelsy in its twentieth-century survival, the coon song. What happens to the "love & theft" of blackface when there is no face?[22] Questions like this one never made it into the debates over copyright, but they

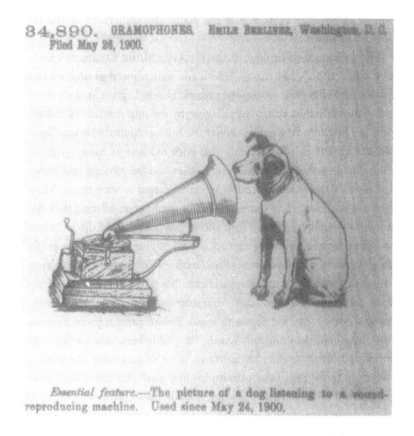

84,890. GRAMOPHONES. Emile Berliner, Washington, D. C.
Filed May 26, 1900.

Essential feature.—The picture of a dog listening to a sound-reproducing machine. Used since May 24, 1900.

comprise the context within which music, musical authorship, and reading music had to make sense. In tracing the legislative history of musical copyright in the pages below, I assume that neither changes to law nor changes to technology can be isolated from contextual and reciprocally changeable parameters of identity and perception. The technology of recorded sound tempered what I call the *visuality of music*, the sum of visual experiences that bolster and accompany musical practice and that extend to the societal norms of visually apprehending racial and other differences. Changing visuality meant changing already complex notions of textuality and performance at a time when other features of twentieth-century music were also taking shape, among them the transnational reach of

American popular culture and the economic structures of the recording industry.

The impassioned listener in the Buckeye Music Company's store
was eager to do what many Americans had wanted to and what a
few had actually done—stop the sadistic Simon Legree. Stowe's novel
had endured half a century of pillaging by melodramatists, parodists,
even the first film companies, when the Edison Manufacturing Company tapped it in 1903. It had also endured just as many years of
piracy by unauthorized publishing houses and translators, and Stowe
had been battered in the courts when she tried to stop them. More
than a few versions had indeed foiled Simon Legree, offering their audiences a happily ending "Tom show." By 1905 and Edison record
no. 8656, Stowe had been dead almost ten years and her copyright
(as insufficient as it had proven) had expired. When Stowe's longtime
neighbor from Hartford, Connecticut, Mark Twain, addressed the
joint congressional committee that convened in 1906 to revise the
copyright code, he had Stowe in mind. Twain came representing authors and had only unkind words for publishers, despite his own
checkered career as one. He spoke in favor of extending the term of
copyright to the life of the author plus fifty years. He saved his humor
for the end of his statement, when he alluded to the arts included in
the proposed law, which extended to cover the mechanical reproduction of sound and images. Twain confessed that he himself had nothing to do with such matters, but he was willing, he said, to support
copyrights for people in those arts, because he felt for them the same
benign interest as a drunkard who, arriving home after a debauch,
finds "his house weaving and weaving and weaving around" before
him, and when, after some struggle, he gains entrance, stumbles up,
and tumbles down the stairs, he exclaims, "'God pity a poor sailor
out at sea on a night like this.'" The pious Stowe might have been
nonplussed by Twain's analogy, but she and her novel had tried to
weather just such storms (Brylawski and Goldman, 4:120–21).[23]

By 1906 the tempest Twain envisioned was particularly intense
for the producers of music and the means of its mechanical reproduction, since records were increasingly a musical form. The Edison

Company persisted with "The Flogging," with vaudevillian dia-
logues and oratory, but "the industry," as it had now become, grew
more and more musically oriented through the 1890s, dominated in
the new century by three patent-holding phonograph companies,
American Graphophone (later consolidated as Columbia), Victor
Talking Machine (later Victor/RCA), and Edison's National Phonograph.[24] With representatives of all three companies in attendance,
most of the congressional hearings of 1906 and 1908 were spent
wrangling over Clause G of the proposed copyright bill, which extended an author's intellectual property to include the rights, "To
make, sell, distribute, or let for hire any device, contrivance, or ap-
pliance especially adapted in any manner whatsoever to reproduce
to the ear the whole or any material part of any work published and
copyrighted after this act shall have gone into effect" (5). Such a
provision would require phonograph record and piano roll manufacturers to obtain licenses from (and pay royalties to) composers,
or to obtain licenses from the big sheet music publishers, who very
frequently obtained the composer's copyright when they published
her or his music.

Copyright had been extended to new media before and both still
photographs and motion pictures offer points of comparison. Photographs were first included in the Copyright Act of 1865. The
courts importantly affirmed their eligibility for copyright in the so-
called "Sarony case" of 1884, on the grounds that photographs "are
representatives of original intellectual conceptions of the author."[25]
The case had involved a "decorative" photograph by Napoleon Sarony of Oscar Wilde, about which—or about whom—the men at the
1906 hearings were still giggling.[26] As Gaines indicates, the Sarony
decision was important in that it defined authorship's ground zero.
The author was merely the "originator" of some work of authorship;
the act of origination became an inference based on the work produced. What this meant by extension "is that Oscar Wilde's own
subjecthood in the photograph secures Sarony's copyright in the photograph" (56, 68–69, 82).[27] The original personality that Wilde displayed in the photograph vouched for the personal originality of

Sarony as an author-photographer. The photograph made the photographer. Wilde responded to this new dynamic in his "Canterville Ghost" (1887). Published after his return to Britain from his American tour and after the Sarony decision, Wilde's comic story has the beleaguered Canterville ghost reduced to "amusing himself by making satirical remarks on the large Saroni [sic] photographs of the United States Minister and his wife, which had now taken the place of the Canterville family pictures" (204). Long a subject in the Canterville house, the manorial ghost finds himself as decisively displaced as the Canterville family. The doodling ghost, Sarony, and the Minister's family all vie for what Gaines calls "subjecthood."

New media did not always inspire new articulations of the author-subject. Motion picture films were deemed eligible for copyright by the courts in *Edison v. Lubin*.[28] The reasoning behind the Lubin decision involved something like Xeno's paradox: because individual frames on a film of Kaiser Wilhelm's yacht were increasingly similar as they were positioned on the film closer and closer together, until adjacent frames were indistinguishable from one another, the film had to be considered a single photograph, not a new entity. Hence it was eligible for protection in accordance with the Sarony precedent and the Act of 1865. Authoring a film was the same as authoring a still photograph. Less straightforward were the bureaucratic mechanics of obtaining copyrights on new representational products. Some film producers scratched out the word "author" on copyright registration forms and substituted the word "proprietor." Film companies such as the Edison Manufacturing Company puzzled over what object to deposit with the Library of Congress in order to register their rights. Literary authors simply sent copies of their printed works (200,000 of them in 1905 alone); photographers like Sarony could deposit a photographic print. Should the Edison studio send a positive print of its films, a photographic negative, a "paper" or "bromide print," or some other object?[29] Did each scene of a film, each new position of the camera, require its own copyright? If, like books, films had to carry a warning label indicating their protected status, how and where should such a label be affixed? Did original film "scenarios," as screenplays were then called, require separate copyright protection? Could scenarios be adaptations of copyrighted works? These details took some time to iron out.

Questions of intellectual property rights proved particularly difficult in the case of music and its mechanical reproduction, in part because of the complexities of what Jacques Attali calls "noise," or the "political economy of music," and in part because the legal standards of intellectual property were written, published works or visually apprehended works the courts could construe as constitutionally protected "writings" in a very broad sense. The combination of pliable uses and new forms made music hard to pin down. The varied economy of American music at the end of the nineteenth century was perched on the edge of mass culture: it relied on noninstitutional as well as institutional means of creating markets for its principal commodity, printed sheet music, while it proved less able to commodify musical performances, phonograph records, and piano rolls in a rational or universal way.[30] It was a heterogenous and multimillion-dollar economy challenged by market shifts, by emergent patterns of consumption, new products, and a changing clientele.

The 1890s song "On the Banks of the Wabash" made its composer Paul Dresser rich when it sold more than 500,000 sheets of published and copyrighted musical notation and lyrics, for which he received a royalty. So popular music was "popular" at least in the sense that people wanted to buy, read, and sing or play it, not because they wanted to listen to it. When they did listen to it, they saw it performed in either a public or a domestic setting. Dresser's song, a ballad with improbably romantic lyrics written with his Naturalist younger brother, Theodore Dreiser, was popular in the additional sense that it partook of a musical tradition distinct from conservatory or "classical" music. Dresser was successful and prolific within the incipient musical culture of Tin Pan Alley. He was what one critic called an "all-round song writer," who wrote everything from coon songs to ballads, though Dresser was recognizably an expert at "so-called 'mother' songs," conservative and schmaltzy ballads; the word "mother" was one he used "in about all his songs."[31] "On the Banks of the Wabash" achieved its popularity in a calculated, if seemingly haphazard, way. According to Theodore Dreiser, 5,000 copies were probably distributed for free in New York City, aimed at singers who might elect to perform the work in public. When performed, free handbills were distributed with the lyrics on them, so that the audience could read along and learn the song, "the sooner [to] hum and whistle it on the streets." Rowdies were hired to sing along or to applaud and cheer wildly in the music hall. Organ grinders were encouraged to play "On the Banks of the Wabash" all over the city. And the publisher manipulated a network of music stores, using discount wholesale/retail agreements and trade advertising to push sales countrywide.[32] A network of music teachers covered the same territories, and various commercial relationships, including "payola," kept the distribution networks in place.[33]

When anyone purchased a copy of the sheet music, it came with permission to perform the work before an audience, understood and every so often made explicit in the form of a notice stamped on the printed score. Phonograph and player piano companies bought one copy of the sheet music, ostensibly rendering their royalty unto Dresser, and manufactured thousands of records and music rolls. Edison's phonograph company issued two different versions of "On the Banks of the Wabash" on wax cylinders, both before the turn of the century, and rerecorded the song on its celluloid "amberol" cylinder, well after Paul Dresser's untimely death in 1906. Composers and their publishers naturally decried the situation; the congressional hearings of 1906 included statements by Victor Herbert and John Philip Sousa arguing for authors' rights. Sousa, a remarkably durable commodity in the bandstand circuit, depended for part of his income, as Dresser did, on the royalties that his compositions earned through sheet music sales. Sousa testified that in using his "copyrighted copy" to make "what they claim is a noncopyrighted copy," the record and music roll companies "take my property" (Brylawski and Goldman, 23). More prescient than most, Sousa also saw the damage phonographs and pianolas were doing to the whole sheet music industry. "You hear these infernal machines going day and night," he said. He also joked that the human vocal chords might soon become vestigial organs and could ultimately disappear forever, casualties of disuse and natural selection. Fewer and fewer people read music; fewer and fewer homes contained musical instruments other than phonographs and "self-playing" pianos. "Popular" music was becoming something people listened to, not something they read, sang, played, or watched. Then Victor Herbert extended Sousa's corporal fancy. The phonograph and music roll companies, he accused, "are reproducing part of our brain" (Brylawski and Goldman, 26). Ironically, Thomas Edison had no quarrel with Herbert's metaphor; phonograph records did indeed "embody" the composer's conception, they bodied forth his idea. It was material embodiment, not the mere conception, that ensured both copyrights and patent rights. Musicians received copyrights when their conceptions were embodied in the printed score. Only that material expression or copytext was protected, Edison argued; as anyone familiar with patent law knew, "If the conception is carried out by a different mechanism they lose their monopoly."[34] But the question remained whether phonograph records and music rolls really did constitute a substantially "different mechanism" as such, or whether they represented some new sort of copy or performance, an unfair use, to be taxed by authors according to their rights.

Battle lines were drawn according to the issue of whether records and music rolls could be construed as copies of "writings" protected by Article One of the Constitution. Writings in this case meant written musical scores, copied and distributed as sheet music. Case law offered a context for the debate. The courts had decided in *White-Smith Music Publishing Co. v. Apollo Co.* that perforated music rolls of a song constituted a single performance, not copies or multiple performances according to the law, so that in buying just one piece of sheet music, the Apollo Company had paid its due. Despite their decisions, the lower courts and then the Supreme Court lamented the letter of the law. In his assenting opinion Justice Holmes chided, "On principle anything that mechanically reproduces the [original] collocation of sounds ought to be held a copy, or if the statute is too narrow ought to be made so by a further act."[35] Rarely was a legislative mandate so clear. The very terms of the court decisions and the ensuing copyright debate reveal a tentative reassessment of reading and writing. Writing was an activity pursued by composers and publishers, not recording artists or record companies, who were only readers. The question at hand was whether the production of records and music rolls created any nonwritten, readable copy. The courts, Congress, composers, and publishers all wanted to sever writing from reading in a new way. By implication they allowed that machinery—phonographs and pianolas—could read.

In the course of the debate representatives of phonograph companies and music roll manufacturers assured members of Congress that their products were not copies of "writings" because they could not be "read," urging no damage to the present law. Frank L. Dyer, Edison's patent attorney, CEO, and sometime-biographer, testified to this effect in 1906 and again in 1908. According to Dyer, Edison himself had once spent many long hours in his laboratory trying to read phonograph records. After recording the letter *a*, "He examined with a microscope each particular indentation and made a drawing of it, so that at the end of two or three days he had what he thought was a picture of the letter '*a*.'" But when he compared two records of the letter *a*, he found that "the two pictures were absolutely dissimilar" (Brylawski and Goldman, 286). Dyer needed to assume that reading was a human activity, not a mechanical one. If even Edison, their illustrious inventor, could not read phonograph records, then they couldn't be read. Dyer argued that what Congress proposed was to copyright sound itself, leaving behind the visual nature of all previous copyrights. By analogy, Dyer asked, why not make it possible to copyright perfumes, extending to the nose the same privilege as the ear and the eye (Brylawski and Goldman, 288)? Dyer insisted that changing the nature of reading meant changing the nature of writing.

Musical culture at large continued to wrestle with similar issues, if not exactly in these terms. The *White-Smith v. Apollo* case had involved the sheet music for a coon song entitled "Little Cotton Dolly," and like the anecdotal account of the Buckeye Music Company, the recorded coon song offers a point of access to some of the more neglected features of the emerging industry. With its increasing diffusion, recorded sound destabilized the connections between hearing music and seeing it performed.[36] "Seeing music" extended to a wide range of social practices, including parlor piano playing, amateur and professional concerts, vaudeville and music hall performances, church singing, and revival meetings. The experienced terms of this destabilization must have differed according to these practices and are notoriously hard to pin down. But the most acute destabilization took place around the recorded coon song, since it was a complex, late-nineteenth-century survival of an already intricate and naggingly visual experience, the midcentury minstrel show. As if a harbinger of all copyright quarrels to come, blackface minstrelsy was rooted in a confusion of origins. Minstrelsy had real and mythic antecedents in the antebellum slave culture of the southern plantation, yet was by definition a northern, urban form. As Eric Lott explains, it functioned in part by offering audiences commodified "blackness" as a way to engage—subliminally and not—the conjunctive class and race politics of the nation.[37] The white construction of minstrelsy's "blackness" possessed inherent contradictions: it played off a contrived sense of authenticity while it also relied upon counterfeiting. The form reenforced racial boundaries by denigrating black Americans, yet it also defiantly transgressed those boundaries for pleasure and profit in what had become marked as a lowbrow, "popular" form of entertainment for the white working class. Minstrelsy subverted the questions of racial essentialism on which it fed, providing a raucous catharsis for matters that seemed so pressing elsewhere in the American national scene: slavery, abolition, and Dred Scott helped form the context and complexion of the minstrel shows; *Plessy v. Ferguson* (1896) would be context for the recorded coon song.[38]

The orientation of blackface minstrelsy was visual and performative, even if a large measure of its dubious authenticity was its supposed appropriation of an oral culture it constructed as "blackness." Seeing a white man with his face smeared in burnt cork was the perceptual and visceral center of blackface for fans and critics alike, though there were black minstrel troops too. Yet just as developments within the music industry tested the persistently visual orientation of copyright law, similar developments long challenged the visual orientation of minstrelsy. One challenge came in the popular, free-standing coon song, performed outside the minstrel show, though well within the minstrel tradition. Another challenge arrived with the player piano and the phonograph. According to the publishers of sheet music, the coon song reached the height of its popularity in the late 1890s, when large numbers of songwriters such as Paul Dresser (who had once been a minstrel) churned out more than six hundred coon songs to cash in on the vogue.[39] By then the immense popularity of minstrelsy had passed; the minstrel show had proved to be an antebellum form that lasted only through Reconstruction and lingered into vaudeville. What this meant is that the sound of white-constructed "blackness" survived without the sight of minstrel blackface, as performers of coon songs could go without burnt cork, particularly as recognizable "coon" elements were incorporated into a variety of different songs and formats. Some unblackened white performers were seen to "sound 'black.'" Finally, when music roll and record companies set out to record coon songs, sounding "black" went colorblind. Whereas minstrelsy had been an acknowledged white, working-class form, the coon song allowed middle-class penetration of its tradition. Coon songs were played in middle-class parlors, concerts, syndicated vaudeville, and the other bourgeois venues where sheet music was increasingly consumed. Class lines were doubly enforced and transgressed in the same manner that racial boundaries were, as middle-class musical practices picked up and dusted off the threads of a working-class form. Like Frank Dyer insisting that the Committee on Patents and Copyrights was trying to protect sound itself, records and music rolls of "Little Cotton Dolly" seemed to assert that white-constructed "blackness" was a matter of sound, not skin color. On the heels of the *Plessy* decision, which had determined "blackness" to be a matter of blood, not skin color, the meaning of music thickened. (The American judiciary deemed white-skinned Homer Plessy black by dint of his African blood.) Now popular culture interrogated music as another possible substance of intrinsic racial difference. Music rolls even seemed to make "sounding 'black'" an instrumental matter more than a vocal one, depending more upon the sound of a piano than the sound of any singer. The same aural associations were affirmed with the contemporary ascendance of syncopation as a black-identified musical feature.[40] Of course the sound of "blackness" was not monolithic; it was never wholly white-constructed and was complicated by other black sounds, by recorded black spirituals, the works of well-known black songwriters and performers, a long tradition of black musical theater, ragtime, the stirrings of jazz, and by the long-standing tradition of ethnic and racist dialect humor in America. In the same month that Edison's phonograph company recorded its third version of Dresser's "On the Banks of the Wabash," its other selections included a "romping coon song" and two "Negro dialect poems," one by the late Paul Lawrence Dunbar.

Earlier record catalogues had been this varied. While major record companies kept white groups on hand for minstrel-influenced burlesques and coon songs, between them Victor and Columbia recorded nearly eighty songs by the African American vocalist Bert Williams, who broke the color barrier at Ziegfeld's *Follies* during his tenure with Columbia. All of this recorded "blackness" without the sight of black, white, or blackened skins was new and uncomfortable, at least for record producers. Furthermore, it comprised an unadmitted counterpart to the legislative debate over separating visible "writings" from the sounds of reading. In both contexts the technology of recorded sound helped to displace the visuality of music. Musical composition, reading, and the creative agency of performance became complicated within new and less visually rooted features of the entertainment industry.

A few months after publishing its anecdote from the Buckeye Music Company, the *Edison Phonograph Monthly* signaled some of the trade's discomfort in another anecdotal compliment submitted to jobbers and dealers, this item under the headline "Mr. Collins Is Not a Negro":

Possibly because of his great success in singing coon and rag-time songs for the Edison Phonograph some people seem to have gained the impression that Arthur Collins is a colored man. Such an impression is naturally amusing to Mr. Collins. It is complimentary, however, to imitate the colored race so closely as to be mistaken for the real article.[41]

This paragraph resuscitates a well-worn trope, one resident in anecdotes about audience members who mistook blackface for blackness. This mistake had been part fulcrum and part safety valve within minstrelsy (as well as imaginably part fact and part fiction) ridden with the racial anxieties, namely regarding supposed risks of racial contiguity, passing, and miscegenation, that it helped diffuse theatrically. Music publishers, apparently fearing the same mistake, had sometimes published minstrel songs with pictures of their blackface performers both in and out of makeup.[42] The *Edison Phonograph Monthly* was doing the same thing in vouching for the distinction between "close" imitation and "the real article." Yet in "Mr. Collins Is Not a Negro," the quickest safety valve proved more elusive than it had before. Unlike the Buckeye Music Company anecdote, this could not be trumpeted as a compliment to the talking machine, only as an "amusing" compliment to the performer Arthur Collins, all because the talking machine had redoubled the problem (part horror, part delight) of identifying "the real article." In this sense race, like racism, differs according to its aural and visual forms.[43] Like Thomas Edison intently trying to discern the letter *a* within the grooves of a record, listeners who tried to discern skin color in Arthur Collins's records were up against something new. Edison had been interrogating the essential nature of records as inscribed texts, while popular audiences were now confronting an incomplete, aural essentialism to the degree that they interrogated records as racialized performances. There was no single, uncomplicated sound for skin color.

It was an interesting historical moment for what is now debated as "black music." Clearly if the crassest technological determinism had pertained, or ever could pertain, the phonograph and then radio would have been colorblind media. But technology does not drive history or culture that hard. The observation that the early recording industry was nondiscriminatory because trade publications seem to have paid black musicians their due[44] needs to include the caveat that the politics of defining and policing racial distinctions has always been a lot more involved than the related now-you-see-it/now-you-don't of discrimination. Paying black musicians their due carried the baggage of needing to tell which musicians were black, in whichever order and for whatever matrix of reasons, good and ill, paying and needing to tell became conscious desires and cultural necessities.

To emphasize the changing visuality of music, phonograph advertisements from the 1890s to the 1920s picture listeners watching the machine. Listeners stare vacantly at unseen and newly reracialized performers, as if by some collective premonition, keeping their gaze steady for radio then television. The gaze itself is oddly communal, fraught with unlikely assumptions about the democratic power of mass media even as it dampens participation. One National Phonograph Company advertisement from 1908 has a mixed-race group of servants staring appreciatively at their employers' phonograph. Below, the caption simultaneously enrolls Western music and the phonograph in the cause of democracy: "One touch of harmony makes the whole world kin." Such rhetoric coincided with Edison's personal expectations for the phonograph, an instrument of social leveling in his ken, since it would allow poor and rural audiences to hear opera. The inventor seems not to have appreciated the anarchic potential of the device as a means for class-crashing or racial ventriloquism.[45] But Edison and many of his contemporaries were sure that they lived in a world of visible certainties when it came to human identity: the inventor interviewed prospective employees while taking notes on the shapes of their heads. And the author Henry James remarked pointedly at Edison's "street boy" face after the two men met in 1911.[46]

A different sort of essentialism was at stake in legislative cham-

"*One touch of harmony makes the whole world kin.*"

THE Phonograph would never have become the great popular entertainer it is but for Edison. He made it desirable by making it good; he made it popular by making it inexpensive.

The EDISON PHONOGRAPH

has brought within reach of all, entertainment which formerly only people of means could afford. It has even displaced more expensive amusements in homes where expense is not considered.

FIGURE 10. "One Touch of Harmony Makes the Whole World Kin." The Edison phonograph addresses class and racial difference in this advertisement (1908).

bers during 1906–1908, where congressmen and witnesses debated how to handle recorded music. Defining and policing authorship, though certainly less inflammatory, proved almost as nettlesome as defining and policing race or class, all because similarly visual habits of definition did not apply as they had before. American copyright law remained dependant upon material forms, so that new forms always caused new problems. And constitutionally protected "writings" were going to be something altogether different if they could be copied out into purely audible forms, without some sort of visible expression. To members of Congress and many other participants in the hearings, it simply seemed intuitive that phonograph records and music rolls—the latter even inscriptions of a sort on *paper*—were copied writings and could be read. As for visible expression, anyone could see the grooves on a record or the holes in a music roll, even if seeing them did not mean anything musically. Witnesses made analogy to hieroglyphics, which resisted reading for a long time, yet were certainly legible. "It is a curious fact," one witness pointed out, "that the earliest known writing, the Assyrian hieroglyphic, was made by an instrumentality very similar to the phonographic needle of to-day impressing itself upon plastic material" (Brylawski and Goldman, 78). Edison would probably have regretted this turn in the debate if he had been following it closely. The analogy to hieroglyphics was unhelpful to his cause, yet he himself had made the same comparison with great satisfaction back in 1888. In an essay entitled "The Perfected Phonograph," which appeared in the *North American Review*, he gloried in his work:

It is curious to reflect that the Assyrians and Babylonians, 2,500 years ago, chose baked clay cylinders inscribed with cuneiform characters, as their medium for perpetuating records; while this recent result of modern science, the phonograph, uses cylinders of wax for a similar purpose, but with the great and progressive difference that our wax cylinders speak for themselves, and will not have to wait dumbly for centuries to be deciphered.[47]

Here the inventor has at once co-opted ancient tradition and assumed the mantle of modern science. The symbolist, nineteenth-century context of his hieroglyph metaphor made using the metaphor in discussions of recorded sound both familiar and extremely powerful. Edison's essay immodestly boosted the inventor to God's place and hinted at the profound centrality of technology in American relations with nature. When the same figure surfaced in the congressional debates over copyright, its appeal was slightly different. Invoking the example of cuneiform meant acknowledging that phonograph records could be read without actually having to read them. The complimentary acts of writing and reading could theoretically be separated by centuries. There was no need to admit so hastily that phonograph records and music rolls could not be deciphered. Meanwhile the traditional object of the hieroglyph metaphor, Nature, jibed well with contemporary appeals to music as a natural, universal language, the spiritual residuum of pre-Babel days. Laden with this metaphorical weight records could be celebrated as what Theodor Adorno would later call "delicately scribbled, utterly illegible writing." The precision and the delicacy with which they were scribbled vouched for the meaning they contained; proof positive emerged from the mouth of a phonograph horn. Members of the joint committees of Congress were only less keenly aware than Adorno would be of the cultural implications of such a mouth.[48] The illegibility of recorded music troubled them as little as the dual nature of patent documents did—both actively concealing the very thing they reveal.

Sensing that the day was lost, opponents of Clause G marshaled every argument they could think of to show the ill-advised, even unconstitutional, nature of the musical copyright provision. Inventors argued that the wording of Clause G directly transgressed their rights "to make, sell, distribute, or let for hire" the devices and processes they had patented. Albert H. Walker, the patent expert who had appeared as an attorney in *White-Smith v. Apollo*, assured the committee that the bill was unconstitutional for a long list of reasons.[49] Many witnesses raised the specter of unfettered monopoly, alleging that the Aeolian Music Company, ever since its early involvement in *White-Smith v. Apollo*, had executed exclusive agreements with almost every publisher of sheet music in America, so that in the event of the copyright bill passing, Aeolian would control all new American music ("a complete monopolistic octopus," [Brylawski and Goldman, 98]). Defending against the accusation that they formed a "phonograph trust" themselves, other witnesses pointed out that composers and sheet music publishers actually benefited from current conditions. Letters were produced to show the way that phonograph companies were solicited by composers and publishers, who sought to stimulate sheet music sales by having records made. A number of witnesses invoked international competitiveness, noting the fact that musical copyright provisions had not succeeded fully in any European country but Italy, where as of March 1908 the matter still had not reached the court of last resort.[50] In France musical copyright provisions had been advocated by composers, only to be overturned in the courts. In England Edison's lawyers had embarrassed one copyright holder in court by asking him under oath if he could understand or read what was on a phonograph record; "He answered, 'Of course not.'"[51] None of the witnesses mentioned Mexico, where all three major American phonograph companies had become embroiled in copyright suits, which had led a participating lawyer for the American Graphophone Company to complain, "It is exceedingly difficult for the American and English mind to foretell how the foreign mind (and particularly a Latin American mind) will work; and it is also difficult to forecast satisfactorily the outcome of litigation in a foreign country."[52] Not surprisingly American record companies were trying to forge or maintain the qualities of statutory authorship abroad with a racialist paternalism not unrelated to their domestic capitalization of coon songs. Lobbyists appealed to Western European models, while developments in Mexico vouched for the newly global and frequently colonialist entertainment economy as well as the precocity of cultural capital in breaching national boundaries.

Even in Western Europe, however, the matter of musical copyright was far from settled. Representatives of the Berne Convention countries met in Berlin during the autumn of 1908 with the mechanical reproduction of music on their agenda. In an early and important instance of such internationalism, Victor Talking Machine, Columbia, and Edison's National Phonograph Companies, all of whom variously possessed shares of European markets and relied upon European composers and performers, joined British and German record companies in trying to stymie any change in the articles of the convention. Paul H. Cromelin, a Columbia executive who had already appeared in the Washington hearings, struggled to draw the three American competitors into cooperation with each other and then with their European rivals. The three American companies urged the secretary of state to press their case, even though America was not a member of the Berne Convention. After this victory Cromelin had limited success coordinating other efforts, and his work behind the scenes in Berlin was to no avail. On November 13, 1908, revised articles of the Berne Convention were signed in Berlin and sent back to member nations for consideration and the emendation of domestic statues. The new Article 13 extended a composer's authorial rights to cover mechanical reproductions. Frank Dyer would testify once more, this time before the copyright committee of the British Parliament, but Britain would go along with the Berne Convention.

The U.S. Copyright Act of 1909 passed a few months after the new Berne Convention; it was signed by outgoing President Theodore Roosevelt and went into effect that July. The earlier Article G, now Article E, protected composers against unlicensed mechanical reproductions. The new law applied only to musical compositions published after the act, so that it enforced an already emerging distinction in the industry between new music (which cost money to record) and old music (which could be recorded for free), with the resulting, ironic split between "the popular domain" and "the popular." From between the two peeked the modern consumer, with changing appetites and mercurial tastes, with resident notions of race, class, gender, and nationality as (some of) the data of culture. Another provision of Article E was called the Compulsory License Clause, which ensured that once a composer licensed one mechanical reproduction, she or he was compelled to license all other proposed mechanical reproductions for a guaranteed royalty fee of two cents per copy. In the abstract, this clause was a reminder that authors' rights in the market were not a matter of natural law as Herbert and Sousa would have it; rather, they were meted out by statute to protect the public interest and, in this case, free-market competition. Practically speaking, this arrangement vitiated the exclusive contracts held by the Aeolian Music Company and added another

The music of Victor Herbert's orchestra under his personal direction will be reproduced exclusively on Edison Records

FIGURE 11. Talent as commodity: Victor Herbert on Edison Records (1909). Victor Herbert's recorded compositions were available to any record company, but Victor Herbert himself would only perform for Edison.

shade of meaning to the "mass" in "mass culture," since mechanical reproduction became in some sense self-perpetuating. When Columbia recorded a song that did well, the other record companies could cut the same record right away. One recording opened the floodgates for multiple recordings as mechanical *re*reproduction followed any potentially successful record or music roll. Compulsory license provided only for musical compositions, not musical performances; the result was an even more heightened sense of "the talent" as a commodity. The record companies intensified their battles for exclusive contracts with celebrated performers. Anyone could record Victor Herbert's compositions after they had been recorded once, but Victor Herbert's Orchestra performed only on Edison records after Herbert signed his exclusive contract with National Phonograph in the summer of 1909. Sousa's band signed an Edison contract one month later. Far from showing Edison's approval of musical copyright,[53] Herbert's contract demonstrates that National Phonograph and other record companies could not afford to be sore losers. They quietly opened composers' royalty accounts in their corporate ledgers.

Clause E had been rewritten to avoid any direct conflict with patent rights, but the two forms of intellectual property were closer in 1909 and 1910 than they had been, or likely ever would be again. The context and the content of the 1906 and 1908 hearings accounted for some of their proximity, as did current commercial practices of tying and price-fixing. The Supreme Court decision in *Leeds and Caitlin v. Victor Talking Machine* was nearly simultaneous with the new act and drew the differing logics of patents and copyrights onto the same plane for the eight years that it stood. The compulsory license provision made musical compositions available to record companies without exclusion; in its *Leeds and Caitlin* decision the Supreme Court regulated what that availability meant, deciding that Victor's patent rights extended to cover which records were played on Victor machines. Leeds and Caitlin, notorious record pirates, had been duplicating Victor records and then competing to supply Victrola owners. The Court now made duplicating illegal by allowing Victor to dictate (to "tie") which records were played on its patented phonographs. So while the compulsory license provision allowed any recording to be remade, *Leeds and Caitlin* assured that remaking would entail live performance, not just "duping" from one record to another. Copyright law made records into special copies of protected writings; patent law briefly protected those copies against duplication. Mechanical reproduction of sound was one thing, mechanical duplication of the same sound was another. Still further contiguity between patent and copyright was established in a decision handed down by Learned Hand early in his career. In 1910 Hand found in *Hein v. Harris* that the copyright for the composition, "The Arab Love Song," had been infringed by "I Think I Hear a Woodpecker Knocking at My Family Tree." Infringement existed, according to Hand, "whether or not the defendant, as he alleges, had never heard the complainant's song, when he wrote his chorus." What this meant was that novelty, not origination or authoring, was the substantial requirement for musical copyright, just as it had always been for patent rights.[54] Sounding the same meant copying in music. This alliance of patents and copyrights was fleeting. In later interpretations of the law, Learned Hand reversed himself, and copyrights diverged from patents accordingly.

In 1909 and 1910 the renegotiated boundary between things and texts had stretched a little thinner. The very lateral groove on a gramophone record, or the up and down engravings of a phonograph needle, were patentable inventions that contained, in some hieroglyphic and as yet undecipherable way, performances of copyrighted sheet music. By extension phonographs and gramophones were "reading machines" more properly than they were "talking machines." For the first time reading aloud was explicitly severed from the human subject. Humans could not read wax cylinders or discs, but machinery could. The reader was less replaced than displaced, pushed aside to make room for the new apparatus that at once shared and complicated human subjectivity. While it is impossible to say for sure how much this new, mechanical kind of reading changed the American experience of reading type or music, mechanical reading did possess some broad implications, both for the emerging culture industries and America's ongoing reconnaissance of the machine in modern life. The legislative construction of reading machines acknowledged continuous reformulation of the "popular" and of popularity in music. For example, the two related skills of reading music and playing an instrument were indeed under assault, as Sousa recognized, in part because reading and playing could now be done mechanically. Mechanical process extended to yet more human functions. In so doing, it continued the often-remarked colonization of the body by the machine. As instruments too, recording phonographs and musical roll perforators could not be authors in the sense that musical composers like Sousa and Herbert were. The machines produced aural copies of writings, performances, and did so until the copyright code was revised again in the 1970s. Then, largely under the pressure of new market conditions surrounding cassette tapes, copyright was extended to cover recordings. Readings under the 1909 Act were made into writings. If human readers seemed newly displaced in 1909, then authorship too had shifted, becoming still more bureaucratic in its relation to the marketplace. Now musical authors had their two cents (literally) riding on every recording, the oversight and collection of which would inspire ASCAP and other societies of authors, along with a torrent of paperwork. The judicial distinction between making mechanical reproductions and duplications further complicated the author's position in relation to the market, by suggesting that the sounds of a composer's work might matter less, in some circumstances, than the means of their production. The distinction between "live" reproductions and other recordings will be examined further in the next chapter.

Nineteenth-century Americans and Europeans had taken phonographs into the jungle and into the arctic. Not only did recording phonographs promise the collection of "live," "native" speech and music but also the reproducing phonograph, as Taussig explains, "proved an easy way for making an intercultural nexus," after dinner, "a new cultural zone . . . for [mutually] discovering strangeness and confirming sameness" (195), as Western travelers and indigenous people could each marvel and be amused at different aspects of their shared experience. American culture, economy, and law in the years around the turn of the century demonstrate that mechanical reproduction at home remained decisively charged with the complexities of that intercultural nexus, a site for participating in experiences of self, identity, and difference. Not only could consumers purchase the recorded hits that "everybody" liked but also they could negotiate difference in the varying cultural valences of Italian opera, "classical" music, "exotic" records from around the world, ethnic records for immigrant niche markets, Simon Legree, coon songs, and burlesques. Differences of class, nation, and race were maintained: phonographs became instruments of "sacralization," helping to distinguish culture as such, and they also became instruments for the maintenance of ethnic identity in the face of assimilationist pressures.[55] The phonograph disrupted identity, in the ways that minstrelsy and popular music had long suggested and in new ways that questioned habits of discerning difference. The technology of recorded sound and the surrounding legal debate challenged existing visual receptor sites within culture, including elaborate practices of experiencing difference. These receptor sites survived in modified form, colored by a new inscrutability, their visual orientation challenged by the prevailing politics of racial essentialism, tested by technologically mediated performances, and, ironically, adapted within a tired nineteenth-century metaphor for nature. America's new hieroglyphics, the visible grooves on a record, were the works of Man and Machine, not God and Nature. The universality that these potent new symbolic actions supposedly possessed did not derive from any transcendental truths or divine omnipotence. Rather they emerged from the questionable universality of music, the penchant of Westerners for lugging their machinery around the globe, and the eager adaptability of cultural capital inserting itself abroad into new markets on every continent. In this spirit, the American phonograph industry established commercial outposts and sent recording engineers to Europe, Asia, and South and Central America.

4

Paperwork and Performance

The U.S. Patent Office's initial designation of Edison's first phonograph as a measuring instrument is a telling point of departure for the exploration of ideas central to the narratives of Foucault, Shapin, Schaffer, and others. These authors have all variously described early-modern science as having purged itself of the human body in favor of what Schaffer calls "self-registering instruments." Experimenters stopped experimenting on themselves and their hapless servants and, instead, began to invoke the objectivity of instruments within a Cartesian polity of professionalized science. The autobiographer dropped out of the scientific article; or, more properly, he learned to cloak himself in an instrumental rhetoric, like the later rhetoric of patents, which was accepted as objective within a community of like-minded male bourgeois. When the phonograph was introduced it harkened back to this purge. At the same time, it showed that the shift in authority from body to instrument, so accomplished in the discourse of professional science, remained relevant and (happily) less determined in Western culture. Shorthand reporters still make history; live performance by musicians is still considered best. Phonograph recordings emerged from the lyceum demonstrations of nineteenth-century culture in much the same way that modern science emerged from the staged experiments of seventeenth- and eighteenth-century natural

148

philosophy. Both were matters of consumption, appropriation, and performance; both negotiated the status of inscribed evidence and the body as cultural norms of knowing and meaning.

In one sense, the authority of the phonograph as an objective instrument seemed assured. The machine fulfilled the unreasoned dreams of shorthand reporters, while accounts like the one in *Scientific American* spoke of an "exact record," or of the "unerring and accurate" reproduction of sound. One of Edison's laboratory employees noted simply that the machine "gives forth the sounds originally spoken," and at least one physics textbook was immediately revised to include an illustration and explanation.[1] Yet when two British scientists did adopt the instrument for the study of acoustics in 1879, they admitted, "The phonograph is in reality a very imperfect speaker, and it requires the aid of much imagination and considerable guessing to follow its reproductions," particularly with regard to "the 'noises' of speech" (Preece and Stroh, 358–59). How could the same device be "exact" and "imperfect" at the same time? Was its authority, the very fact of its performance, to be as pliable as the accuracy of shorthand, with its generous semantics of "verbatim" reports? Questions like these are particularly difficult to approach given the burdens of hindsight. Even listening to a replica of the 1877 phonograph, while confirming the designation of "imperfect speaker," does nothing to explain "exact," "unerring," and "accurate." One is left with the sense that people's bodies, specifically their ears, must have been different in 1877 than they are today. In this way contemporary understandings of any past technology, on its own terms, may not be easy, since understanding itself is determined by present experiences of related technology. Particularly when the past technology in question is now called *media*, it seems certain that its meaning is *mediated* in anachronistic ways, precisely because the media is so powerful in its creation of the present. One doesn't just go to the movies, surf the Net, or listen to recorded sound, one knows them and has learned to understand them in a host of tacit, visceral, and unselfconsciously seductive ways. What I am proposing, in other words, is that one can go out into the woods and experience what life was like before the incandescent lamp, electric light and power distribution, and automobiles, but no matter how far into the woods one goes, one cannot experience what life was like before broadcasting, for instance, or before recorded sound. Like language itself, there is some level at which media help "wire" people for the thinking they do.[2]

This chapter looks at evidential claims made on behalf of phonograph records and early motion pictures as a means of recovering some of the experiences of listening to recordings and looking at films at the end of the nineteenth century. Certainly, examining the rhetorical formulations of early-modern science has helped historians of science and sociologists of knowledge discern the textual, bodily, and otherwise social construction of modern, objective inquiry. Likewise, looking at the ways that phonograph records and films were identified as mimetic products can help untangle what they meant as mimetic experiences. Of course, the human ear was anatomically the same in 1877 as it is today. What needs to be considered is how the new experience of listening to recorded sounds was constructed socially. Records had to be understood first as literal reproductions, then as "cultural" ones. The preceding chapters have hinted at a few of the broader aspects of these constructions, such as the experienced textuality of phonography and the phonograph, complications and displacements of visuality, emerging forms of business organization, and deep-seated assumptions about the social and psychological integration of technology. This chapter turns with greater resolution to what Martha Banta calls "the narrative productions that entered the marketplace" in her discussion of the culture of scientific management (5). Inventors, investors, manufacturers, and salesmen all identify the products they think will make money, and do so in reaction to perceived conditions and changes in the marketplace. They narrate (however accurately or inaccurately) both what their products are and what they should mean to consumers, narrating—as in the case of phonograph records and films—the desirable characteristics of inscription and representation as cultural commodities.

As Chapter 3 shows, the patent process characterizes itself as a paper bridge from the laboratory to the marketplace. Yet what is in one sense a bridge is in another only the most formal discursive element in a disordered paper conduit, a shuffle of paperwork that extends at least from the proverbial drawing board to the gummed "patent pending" label and other commercial marks affixed to new products. When they travel beyond the laboratory walls, experimental notebooks have a particularly privileged status as inscriptions. Edison frequently produced his notebooks in the courtroom and in patent office hearings. Yet research notes were not the only, or even the usual, form of paperwork to emerge from the inventor's laboratory quadrangle. Patent applications made their way to Washington; journalistic descriptions made it into the daily papers; and huge masses of personal, business, and legal correspondence ventured out to characterize the technological work that was taking place at the lab. At the end of this long road of paper was the product label and the literature of advertisement and promotion.

The label is a vital cultural nexus, a point where producers meet consumers, where owners meet spectators, where novelty and originality enter the commonplace of the market and commodities perform. Patented products (and copyright ones) are, by law, labeled as such. But the product label does more than identify proprietary rights: it brands, distinguishes, and is aggressively tailored to the form of the product.[3] Not only are the labels on small things small, the labels on cylindrical things cylindrical, but also labels necessarily take account of the product's intended ontology, of what the product *is*, frequently by specifying its origin, composition, and use. As such, labels are but the simplest inflections within multiple and difficult narratives of knowledge and power. Like patent documents, labels mark the accomplishment of research in an arena called "the market," rather than law. Labels, like patents, are descriptive within a consensual discourse; they make sense by responding to and publicly joining similar expressive acts. The sense they make is in hinting broadly at narratives of production and consumption, substance and sales, "technology transfer" from laboratory to marketplace. However tenuous it may appear at first to propose labels as a discrete nonliterary genre, this chapter assays the consensual features of early phonograph and film labels as such. Here, I explore some of the connections between paperwork and performance that helped make these new cultural forms meaningful, though they were themselves neither paper nor performed live.

Technology transfer is a misnomer, even according to the scholars who coined and elaborated the term. The word *transfer* suggests a material relocation, as if getting the phonograph out of Edison's laboratory and into the American living room was simply a matter of grabbing hold and moving the machine. But this "transfer" means a lot more; it denotes a whole matrix of transactions buffered by technological imperatives (a viable phonograph and record) and informed by expectation, rhetoric, taste, communication flow, and economic conditions.[4] Clearly preconceptions about the phonograph as a measuring instrument or a language machine helped determine its early identity as a product. Preconceptions matter in part because they inform, test, even debilitate technology transfer. The early local phonograph companies failed because they tried to effect a technology transfer from lab to market using the wrong narrative of how the phonograph should be used. Having assumed the wrong narrative, they lacked the entrepreneurial skill, economic resources, contractual rights, and access to production needed to revise their narrative appropriately. More simply, they failed to hear the other narrative, the one about mechanical amusement, that consumers found made sense. Consumer feedback is important but notoriously hard to isolate and study. Often the responses of flexible producers, who relabel products and revise their narratives in order to succeed, prove the best available guide to consumer feedback.[5]

Product labels denote reification. The "real thing" does not generally need a label. A tree in the forest bears no label; only when it is identified as timber, cut, and processed as lumber does it need to be tagged. In the abstract, this is the nature/culture distinction. Nature *exists*, essentially, while culture labels, co-opts, and transforms. To the extent that a tree in the forest does possess a name like *acer* or *quercus*, maple or oak, nature has already been appropriated into a taxonomy and thus into culture. By extension, cultural products are more and more reified as they require further layers of labeling. The marketplace requires that tagged, processed lumber be given a grade and, ultimately, a price. While labeling is thus a cultural and not a modern phenomenon per se, modern enterprise and managerial practice, in general, and mass production and mechanical reproduction, in particular, have clearly added layers to the conduct of labeling. More products exist to be labeled, and more layers of reference have been required by additions to managerial and other bureaucracies.

The most ironic effect of modern accelerations of labeling is that "nature" has become one of Western culture's most powerful labels. Beginning after the mid-nineteenth century, products became particularly valued if they could be characterized as "real," "genuine," or "natural." With increased use, labels like these became strikingly more pliable in their connotations. Not surprisingly, the new value that American culture placed on "real" things coincided with the substitution of multiple machine-made products for ones that had previously been made by hand. That this condition of modernity should possess irony must be laid at the door of modernism and its antecedents, which, as a loosely designated chorale of style and artistic movements, prattered against the complacence of bourgeois culture, with its dependence upon imitations and its faddish affection for acts of simulation and gestures of authenticity. This reactive irony, the double consciousness of modernism, evinces an awareness of labeling and its acceleration, if judging only from avant-garde attempts to assert mere referents as emphatically and singularly authentic ("A rose is a rose is a rose") and to complicate or disavow the symbolic altogether, as in the drive toward abstraction in the fine arts. In both the culture of imitation and the culture of authenticity that Miles Orvell describes, "the real thing" stands in distinction from the realistic, and the operations of that same distinction locate and deeply color the experiences of representation and the appropriation of new mimetic forms.[6] It is not the difference between the real and the realistic that matters so much as the cultural operation of that distinction in the location and experience of representation. The great satisfaction of the blind taste-tests between colas that advertisers ran in the 1970s was not that "The Real Thing" tastes any better or worse, but that the colas taste different at all.

Phonograph records and motion picture films emerged as amusements at different ends of the 1890s, doubly figured as realist vehicles and fantasy machines, contradictory reservoirs of reality and realism. Despite many obvious differences between the two forms, their status as representations shared certain general, historical features. They were "unexpected sources of pleasure" (Musser, *Emergence of Cinema*, 494), nonhuman, objective instruments, but also "just" entertainment, at once decadent and banal, useless and diverting. They seemed to give uncannily factual representations of reality, and then were used to offer the universal "truths" of music and melodrama: fact versus truth, objectivity versus subject. These complications were played out, in the case of the phonograph, largely as a matter of metaphor and personification, the mechanical emulation of human performance. As a language machine the phonograph performed the acts of talking and reading. As an amusement device it continued to personify mediations between machinery and human experience. In early films, meanwhile, the peculiar confluence of fact and truth was evident in the juxtaposition of a powerful documentarian impulse beside the prurience of vaudeville and, to a lesser degree, the magic-realism of Georges Méliès and his ilk. Reality and fantasy seemed equally available to producers, who filmed both civic and theatrical events, scenery and news, as well as "acts" and "gags." Hints of the same refractions had certainly surfaced before, in debates concerning the aesthetic and/or literal status of photography, for instance, and contemporary debates concerning the "yellowness" of journalism. Now the forum for the debate was a changing market within which phonograph and film companies competed for and won mass audiences. Competing and winning had first to do with the reality claims made on behalf of early motion pictures and phonograph records, with the status of film and record as mimetic reservoirs, products of self-registering cameras and phonographs. One program distributed for an evening of short film subjects, entitled "Edison's Startling Invention: Life Motion Pictures" (ca. 1902), captures the confusion.[7] The motion picture program, a bit of obsolete paperwork by today's standards, explains, "If your own family photograph were taken by this new process you would appear alive and full size on the canvas . . . laughing and shaking hands as if really alive." Further, the presentation "will be shown on the immense canvas in real life movements so natural you would feel that horses and men would actually leave the canvas and come dashing into the audience." The promoter's emphasis on life, the reiteration of "alive," "really alive," and "real life" within the space of several sentences (modified by "so natural" and "actually"), identifies the power such labels possessed while it also reveals an unadmitted concern about the documentary status of motion pictures as representational products. The promoter's emphasis on the size of the moving images ("full size," "immense") indicates the recent shift in the development of motion pictures, away from the cabinet technology of the peephole kinetoscope with its "parlor" market, toward the theater technology of projected film, with its developing nickelodeon market.

In a certain sense film programs were a quickly abandoned attempt to label individual film performances; film producers felt a fleeting need to frame film presentations with a distributed bit of printed material, the way that live theater was framed. But new technological developments stimulated new forms of labeling. Many kinds of labels were tried, modified, further modified to suit the perceived technological and commercial realities at hand, and then sometimes abandoned. Like "Life Motion Pictures," early phonograph records were labeled by a pointed distribution of printed material. For example, when the first five-inch Berliner disc records were manufactured in Europe in 1889, "the sound quality was so dubious that a small rectangular paper label imprinted with the actual words was glued to the back (even for *The Lord's Prayer!*)" (Koenigsberg, *Patent*, 56). This label was tailored to frame the recorded performance, giving the listener a little paperwork to make sense of the experience. When record quality improved, this sort of label became obsolete, and when later labels on record boxes or jacket sleeves included printed lyrics, those labels were tailored to a slightly different purpose, indicating a further negotiation of what the listening experience entailed. In this way phonograph record labels and motion picture labels developed in roughly parallel fashions. Just as legislators, manufacturers, and judges needed to figure out how these new forms related to intellectual property concerns, the producers, manufacturers, and marketing agents of phonograph records and motion pictures needed to figure out what labeling forms suited their new products. Each labeling innovation was the result of an awareness of what the product was, the consumers' awareness often glimpsed through the producers' decisions. Some labels proved to be mistakes, others were quickly obsolete or unnecessary. The comparable developments in record and film labeling provide an outline of the way in which these products of mechanical reproduction became both apprehended as cultural artifacts and constructed as cultural commodities. In particular they register the ways in which phonographs and films emerged from the inventors' workshops and into the marketplace, where contemporary notions of performance underwent necessary interrogation and where performers' bodies became newly mystified in the mechanical productions of new cultural forms. Qualified "live" performance on records and films was the newest feature of culture in the fin de siècle, and labels offer one way to trace the features of such qualification.

Internal labels were the most unique form of identification imposed upon early phonograph records and motion pictures. Because they were inscriptive products, recordings and films could possess labels of a comparably representational nature: recordings could label recordings; films could label films. As one magazine reported of the phonograph in 1891, "Before the piece is recorded, the title is shouted into the machine" (*La Nature*, cited in Attali, 91). This form of internal label, though infrequently remarked upon, survived well into the twentieth century. Records produced by Edison's National Phonograph Company began with a recorded announcement,

"Edison record," followed or prefaced by the name of the performer or performers and title of the piece that followed. These announcements introduced recordings until January 1, 1909, when they were discontinued after a lengthy debate among Edison and his executive officers and after some polling of phonograph distributors and members of the public.[8] The reasons for these recorded labels were enumerated in the debate. "I have always argued in favor of retaining the announcement for two reasons," wrote one of Edison's lawyers; "1st. For legal reasons, in order to give us a stronger position in case of unlawful duplication; 2nd. For advertising purposes."[9] His "legal reasons" suggest the ongoing debate concerning musical copyright, the untested character of records as copies, and the rigorous patent litigation underway. His "advertising purposes" reveal the extent to which phonographs were still listened to outside the home and as an "instrument of sociality" (Attali, 69); they were played in public or private social gatherings, in retail establishments, and in a few remaining phonograph parlors, where internal labels made it possible for listeners to identify the records they heard and later make their own purchases based on that experience. Some of the information provided by the announcement was provided in no other way *on* the record itself, only on its packaging and attendant sales literature, even though by 1908 the National Phonograph Company was engraving the title of each recording around the rim of the cylindrical record.

In a memorandum to Edison, Frank L. Dyer made three persuasive arguments for the discontinuance of the internal label. First, omitting the announcement would save space on the record surface, making longer selections possible—length was a heated issue in 1908, and the National Phonograph Company had just introduced its own four-minute record, the "amberol," to compete with other long-playing records on the market. (Length would remain important for decades, reaching its vinyl apotheosis in the aptly named "LP"—long playing.) Second and "most important" to Dyer, "by omitting the announcement the foreign records [with foreign-language announcements] become immediately available for sale in this country." Dyer shrewdly wanted flexibility to be a feature of the company's pursuit of global markets. And third, Dyer confessed, the company was receiving letters from customers, who wrote that they "object to the announcement preceding the selection as detracting from the selection." The internal label, seemingly unnoticed for twenty years, had invaded the representation it sought to identify. Something seems to have changed in the way people considered and listened to records, which made the announcements intrusive. It was thought best to drop internal labels in favor of more external forms of labeling, more paperwork.

The development of motion pictures involved a flirtation with internal labels that has proved longer lived. The earliest films frequently had no titles; distributors and exhibitors were left to call them by whatever titles they wished. But in the mid- and late 1890s moviemakers shot scenes in which the initials or insignia of their companies were featured prominently. One frequently reproduced still of *Annabelle Butterfly Dance* (1895), for instance, contains Annabelle and, resting at her feet, a large capital C, standing for the Continental Commerce Company, which distributed the film. Annabelle is dancing and the C is resting, its immobile presence conflicting the subject of the film, which modulates distractedly between Annabelle's performance and the Commerce Company's production. Later, company initials or names were moved to the title frames where they were repeated whenever the silent film gave scene titles or dialogue (two specialized forms of internal labeling). Biograph's B entwined like ivy at the corners of the title frames in films like D. W. Griffith's *A Corner in Wheat* (1909). Like *Annabelle Butterfly Dance*, Griffith's film possesses a conflicted subject; its focus modulates between the film's narrative (corner in wheat) and the film's production (B in the corner). Both of these forms of alphabetic labeling did not last. What still remains is the internal labeling of title and credits at one or both ends of the film. Probably because films could not be handled physically by individual users the way that phonograph records could, internal labels remained more viable than external ones, though internal labels were quickly banished from the

FIGURE 12. Internal labels and the conflicted subject of early films. "Annabelle Butterfly Dance" (1895). The "C" is for the Continental Commerce Company. (Courtesy of Ray Phillips)

middles of dramatic scenes and pushed to the first and last few feet of the reel. The differing accommodation of internal labels in phonograph records and motion pictures suggests that a distinction homologous to the one between producer and consumer goods has long existed in the entertainment economy, a distinction between unhandled and handled commodities, between those more or less rooted in performance. Records are played while films are shown, a distinction that measures the umbilical distance from producer to product in strangely tactile terms, and one that has blurred with time, challenged by radio broadcasting and video cassettes. The radio un-handed phonograph records, while the VCR let audiences touch the film.

One way to understand the phenomenon of internal labeling might be to think in terms of books and other text-objects; printed books bear printed labels. Yet the internal/external distinction hardly pertains. Even legally the comparison has been squelched. In what is still called "The Mark Twain Case," Twain sued a publisher who was selling a collection of his uncopyrighted works. Unable to claim copyright infringement because he hadn't bothered to copyright these short works, the author sought to bar the use of "Mark Twain" as a label. But the courts ultimately ruled that authors are entitled to no protection in addition to copyright. In other words, the name "Mark Twain" could not be considered a brand name for merchandise in this case, even though books were a hotly contested commodity at the time, amid ongoing changes in market structure and rampant piracy. Instead, "Mark Twain" was a just a pseudonym, investing its user with only as many rights as the name Samuel L. Clemens. At least because they are authored, books are not the same as other commercial products.[10] The irony would not have been lost on Twain that if the pirates had been printing a book he had *not* written and passing it off as his, then the law would have afforded him some protection. In Britain Lord Byron had managed to stop just such an edition of verse, unscrupulously promoted as "by" Lord Byron, though other authors were less successful at stopping similar frauds.[11] The "Mark Twain Case" in one sense denied the exteriority of a book label. The words "Mark Twain" on the cover or title page of a book were not a proprietary label per se, only the reiteration (or not) of copyright privileges inhering to the creative work.

This was not the case with the words "Thomas Edison" on a projected strip of celluloid. Edison was able to sue pirates who duped his films on the grounds that they had, in the process, duped his trademark signature. It wasn't the most effective way to stymie the pirates, but it seemed available because the authorship of films for a time appeared so much more obscure than the authorship of books. Like Clemens, Edison busied himself and his lawyers carving privacy rights out of his own publicity. Edison applied for and received a trademark for his signature, protecting it for use in selling "certain named scientific and philosophical apparatus."[12] He and his lawyers also engaged in hundreds of legal actions to enjoin the use of the name "Edison" in any form, pursuing everything from Edison patent medicines to Edison corsets and cigars, arguing in effect, that Edison's acknowledged reputation as a technological expert gave him enlarged rights to use the name. He even squelched use of the name by his son, Thomas Edison, Jr., when his son embarked on his own commercial enterprises, resorting to the courts on more than one occasion. (Tom Jr. subsequently adopted the alias Burton Willard as he spiraled into alcoholism.)

Of more direct relation than printed books or Edison corsets to early internal film and phonograph labels was the experience of the inventor in the laboratory. Internal labels were a holdover from the lab, where the inscriptions necessary to the invention of inscriptive technologies could be marshaled to provide evidence of their own existence. Experiments with phonograph recordings were noted in laboratory notebooks and also inscribed *on* phonograph recordings. Experiments with film technology were necessarily inscribed on film as well as on paper. In Edison's laboratory the self-registering instruments of the scientist comprised an end as well as a means of inquiry. This simultaneity possessed a number of important implications regarding the emergence of records and films as cultural forms. The moment when an inscription bears evidence of itself is the hallmark of inventing representational technologies. Mark Twain, for instance, sunk a lot of money into the invention of a typesetting machine and documented its (fleeting) success in his personal notebook by pasting in a celebratory page composed on the typesetter itself.[13] Similarly, the experimental notebooks that document Edison's early development of the "electric pen" stenciling device are full of stencils. Over and over again Edison and his associates try versions of the phrase, "Now is the winter of our discontent made glorious summer by this son of York." Each new version of the phrase, each new version of the electric pen, represented a small victory in an expected line of succession as each improved prototype got closer and closer to a market-ready version of the stenciling equipment. Product eventually eclipsed invention. Edison selected Shakespeare randomly perhaps, electing the first line of his favorite play out of myriad possibilities, but the phrase appropriately became his "Eureka!" implying a nowness, an emphatic present, as stencils made with the electric pen got better and better as the device was improved.[14] Even in the patent caveat filed for the device, the words "now is the win" are visible in one illustration.[15] These self-evidencing inscriptions were echoed in the self-identifying nature of internally labeled records and films.

Like internal labels, external product labels are not without tap roots in the laboratory. Inventors necessarily identify what it is they invent, though an inventor's label possesses few special rights as the "correct" or eventual name for a technology, a name that must be negotiated beyond the laboratory walls and over time. Edison's neologism "phonograph" lost out in Britain and stuck in the American market, though not without contests from "gramophone," "graphophone," and "talking machine," as well as later incursions by the trade name "Victrola." (The label "phonograph record" seems to have slipped into colloquial discourse unnoticed and uncontested; the term *record player* became a neutral, noncommercial synonym for "phonograph.") Stakes are high. The reputation of the inventor, the invention, and—with the addition of trademark status—a substantial and exclusive advantage in the marketplace can all ride on such negotiations. These stakes explain something of Edison's tone in a letter to the editor of the *Chemical News* of 1878, in which he claims that the telephone relay lately publicized by E. J. Houston of Philadelphia is really a "speaking telegraph" that he himself had invented the year before. Edison writes:

[Houston's] statement that a "speaking telegraph" is not an articulating telephone is a pretty fine distinction without a difference. Change of form and name appears to be an easy and favorite method now-a-days of making discoveries and inventions.[16]

Edison asserts his priority by seeking to control the labeling of the invention in question, making the point that two different names do not always denote two different devices. The implication that neither "name" nor "form" makes an invention is in keeping with the logic of both idea letters and patents, which hold the idea, a bit of mental topography, as the stuff of invention. Edison's sarcasm is a hasty, rhetorical contribution to negotiations already underway. Like the nomination of the phonograph, naming the telephone relay depended upon other technicians, publishers and publicists, users and consumers, as well as competing inventors, with their self-promotional machinations. Edison knew "A *name is very important*," that an inventor's name for an invention had a greater chance for success if it was well thought out. In naming his talking-picture machine the "kinetophone," he chided his executives with characteristic bluntness, "Before you finally decide let somebody that got [*sic*] a little more imagination than a mule make a list of 20 or 30 names."[17]

The connections between the laboratory and the marketplace were never more explicit than they were on Edison's product labels. Recognizing and encouraging the weight of his own celebrity as an inventor, Edison plastered himself and his lab all over the products he offered for sale. Records were "Made at the Edison Laboratory, Orange, New Jersey," a claim that elided existing corporate, personnel, and financial distinctions between Edison's experimental and commercial enterprises in West Orange. Records were presented as if they were the individual inventions of the lab, rather than the bulk products of the Edison Phonograph Works and the National Phonograph Company. (The implication ill-served them later on, as market emphasis continued to shift from the novel to the fashionable, from the invented to the up-to-date, and Edison's cylinders started to look quaint rather than modern.) Not only the "Edison" brand name but also the inventor's trademark signature appeared on products. On record labels the inventor's copyright portrait appeared engraved in an oval roughly the same size and proportion as George

FIGURE 13. Echo all over: Edison cylinder record label. The company had offices in the vaudeville district in New York, but the records were supposedly made "at the Edison Laboratory."

Washington in his oval on the dollar bill. Edison was clearly a secondary commodity in the sale of his products; purchasers were buying Edison at the same time that they bought a phonograph or a record. His person merged with his products. The merger was evidenced lexically in the 1910s, when his dictaphone was dubbed the "ediphone," and in the 1920s, when his line of tabletop appliances was dubbed "edicraft." The poet Hart Crane coined his sardonic term, "Ediford," with reason, evoking America's two idiosyncratic technicians at once in a section of *The Bridge* that begins with the hail, "Stick you patent name on a signboard / brother—all over" (16).[18] The confusing number and variety of "Edisons" labeling any phonograph record made the National Phonograph Company sales slogan, "Echo all over the world," seem ironic. Records bore the Edison brand name, frequently featured the Edison Band, were recorded "at the Edison Laboratory," protected by Edison patents, and marked with Edison's signature and portrait. The Edison echo was deafening. Edison's competitors were far less interested in marking the connections between laboratory and marketplace. If they relied upon secondary, celebrity commodities, then it would be the "star" they promoted[19]—Enrico Caruso in the case of the Victor Talking Machine Company, and a little later, the "Biograph Girl" in motion pictures by the production company of that name.

Another secondary commodity proved to be packaging: the cardboard tube that cylinder records came in; the oblong box for piano rolls; the folded pasteboard cartons and containers for products like breakfast cereal and milk. Importantly, such packages had to act as reservoirs for attendant labeling. Though perhaps consumers did not recognize packaging as a commodity with the clarity that manufacturers had to, packaging became a potential means of market control. Manufacturers patented their boxes and cartons, attempting to dominate the market in their products by obtaining a monopoly on the way the product was packaged. New processes for labeling were also patented. Box making, label, catalogue, and other printing became secondary enterprises for many manufacturers, a necessary jobbed-out, contractual expense for others. The National Phonograph Company entered into dubious contracts with the Seely Tube and Box Company and the Essex Press of Newark to meet its needs, although the latter may have turned out to be an extortionary enterprise controlled by moonlighting Edison employees engaged in conflicts of interest.[20] In their commodification of record packaging, phonograph companies were acceding to commercial and cultural trends. American commerce was moving away from the sale of raw goods in rough hands toward the sale of processed, packaged goods distributed through a prearranged hierarchy of wholesalers and retailers. Phonograph cylinders were new products par excellence. In addition to requiring elaborate retail demonstrations, they required a new, distinctive packaging and even required a new sort of handling. In its sales literature the National Phonograph Company instructed its consumers in the proper way of holding a cylinder, protecting the record grooves, spreading the index and second finger against the inside of the cylinder.

The most striking thing about the labeling of phonograph cylinders is the great profusion of label forms and contents. It is tempting to see this profusion as an indicator of an unsettled technology, of a product that was still hotly contested in patent disputes, copyright legislation, and extreme competition from pirates and price cutters as well as from the manufacturers of the disc record, which would later be the "winning" form of record in the contest for market dominance. It is equally tempting to see the confused profusion of early cylinder labeling as an indicator of emerging cultural forms, of the newly figured commercial status of amusement, on the one hand, and the accompanying proliferation of mechanical reproductions, on the other. Without discounting either of these temptations, clearly part of the confusion in labeling cylinder records stemmed from their lack of actual surface area, which was a technological imperative. Labels could not be put on the outside of the cylinder, where the record grooves were. That left only the two ends of the cylinder. In the 1890s some companies molded their company names into one rim; in 1892 Edison pressed paper rings onto the rim, giving the recording artist and selection title; in 1900 the Columbia Phonograph Company put selection numbers and titles at the extreme margin of the recording surface; and in 1904 the Edison company began to engrave selection titles along one rim.[21] Later records surrounded the record title with still more information engraved on the rim, including the recording artist (often abbreviated), as well as a tiny reproduction of Edison's signature and the abbreviation "Pat'd." Given the limited amount of space available, the preference for indexical labels, which change with every selection, rather than proprietary labels, which boast company names, was an important one. Like the later discontinuance of internal labels, it suggests a growing awareness of individual selections as the central, inviolable products being sold in the new entertainment economy; the title of a recorded selection identified what was most important about a record.

In 1903 Eldridge R. Johnson of the Victor Talking Machine Company received a patent on the circular, central label for discs, which have more available surface area.[22] His patent provided for "flush mounting or a slightly recessed area to protect the paper insert, which was applied during the stamping process when the record was soft and tacky,"[23] and his paper labels became and remained the industry standard, providing room for all the necessary indexical and proprietary information. Despite experimentation with picture records (labeled under a transparent recording surface) the makers of cylinder records were never successful in providing enough surface area to fully label their records *on* the record itself; they were obliged to label records on their packaging. As a kind of compromise position, Edison's phonograph companies took to using paper insert slips resting inside cylinders, inside their tubular boxes, as well as labels on the outside of the boxes. In the 1890s these insert slips (roughly 4" x 2") contained mostly indexical information, sometimes even handwritten or hand stamped, but after 1900 they became full-fledged printed labels, with indexical and proprietary information, including the usual portrait of Edison. These printed slips included the selection title, catalogue number, and the words "Genuine Edison Record" in a circle that could be cut out. In small letters the circumference was marked "The circle fits top of pasteboard spindle," encouraging consumers or distributors to enter into the labeling process themselves by cutting out the circle and attaching it to the box top or to the spindle in a specially designed cabinet for storing cylindrical records.[24] Later box tops were factory labeled, and the record inserts were replaced by record "slips," paper liners wrapped around the cylinder inside its tube. These liner notes contained information about the performer or composer, gave the lyrics, or added some other bit of explanatory detail to the indexical labels already provided.[25] Not only did consumers have to handle these new products in new ways, but handling them was attended by a variety of paperwork, sheets to be removed, pasted, unwrapped, separated, and read, all so that the consumer might join Edison as his company in the authorized identification of consumable goods.

As excessive as this cylinder labeling seems, with internal announcements, engraved rims, printed box labels, insert slips, liner notes, and cutout spindle tops, such labeling was a corporate response to what cylinder records were perceived to be as cultural artifacts and commodities. Records were slippery; they *needed* that many labels. The experienced ontology of the cylinder was rendered less obliquely in the content than in the form of these various labels.

Edison labels emphatically registered the laboratory origins of his product, for instance, keeping the act of invention as present as possible in the corresponding acts of manufacture and consumption. Other proprietary information is registered at multiple levels. The record company, brand name, and trade name are given side by side with notice of patents on recording technology, record material, duplicating, packaging, and labeling processes. More than ten patents are referred to by date on some Edison cylinders, along with the threatening notice of "other patents pending." At least two trademarks and two copyrights protect most Edison labels, covering Edison's signature and portrait, as well as trade names and other features of the label. Indexical information includes the selection's title, author, performer, and accompanyist, but extends as well to a variety of index numbers imposed in the process of manufacture. Each selection had a catalogue number, assigned to identify it in the company's stock and sales literature. Additionally, each amberol cylinder was impressed with two or three tiny numbers and a series of one to four pin holes. The numbers alluded to the quantity of copies made from one of a series of molds, while the pin holes indicated the different "takes" of a recording session.[26] This sort of indexical labeling shows modern production methods, in general, and mechanical reproduction, in particular, to have accelerated the function of labeling. Making its best effort to produce large quantities of the same records of the same selection, the National Phonograph Company nevertheless distinguished its products from one another. Record and film companies were caught in a necessary tension between similarity and difference in which the realism of their recordings was interrogated by necessary, slight distinctions made between multiple records of "the same" mechanically reproduced selection.

These index numbers provide a reminder that recording, like duplication, is in some sense a complicated form of labeling. Recording and labeling share related mimetic claims. When Edison recorded a particular selection, he was labeling a live performance, inscribing it, distinguishing it, appropriating it into a new, reified existence. Similarly, when a nineteenth-century scrivener, a mimeograph, or photocopy machine copies a page, each word of the original page gets labeled with a copy of itself. What is produced is now called a *simulacrum*, like a map that is the same size as the country it maps, or like a globe the size of the earth.[27] It is possible to imagine cases in which such an ornate labeling of the real thing abrogates its realness. If duplicates are really exact duplicates, then originality becomes unlocatable and irrelevant, as Walter Benjamin realized. Realistically, this extreme is approached asymptotically, but never reached. A scrivener's copy gets stashed in a filing cabinet as a copy, never losing its status as such. Twentieth-century Edison records were molded from a "master" made from live performance. They reproduced the original music reproduced on the master, but never became either of these realities, as much as they might mimic, with announcements, canned applause, or other devices. If Walter Benjamin's concept of *aura* is what distinguishes an original work, then it is *quality* that denotes a reproduction or a duplication. Records and duplicates always possess quality in the sense of faithfulness; copies can be "true" copies or "near" copies. Reproductions can be good, bad, or mediocre. They can be better, worse, or equal to one another, but they do not lose the *aura*lessness of their reproductive purpose, though attention to the fact may be deflected by advertising and promotions. Mark Twain makes this point backhandedly in an autobiographical essay entitled "The First Writing-Machines." Using one of the first-marketed typewriters, Twain responded to a particularly demanding request for an autograph letter by typing the letter, signature and all. The gap between holograph and typescript would not by itself have been inherently dissatisfying to the autograph seeker (Edward Bok in this case); rather the supposed lack of qualitative distinction between one typescript and another causes the trouble. There is nothing personal about the form of a typescript. For this reason, the typewriter was disparaged as an impolite, businesslike means of correspondence for many years. The National Phonograph Company relied upon this same supposed lack of qualitative distinction, the same lack of personality, in issuing large quantities of "the same" recorded selection, at the same time that it necessarily indexed individual records as different. Personality can be hard to stem, however, and it is to the additional subjectivities of mechanical amusement that the rest of this chapter turns.

Although records were slippery commodities, requiring multiple labels to function in the marketplace, they were also admittedly slippery products, requiring another set of labels to function coherently in the process of their own manufacture, allowing producers to "keep track." Different labels play to different readers; only National Phonograph Company initiates could read the pin holes in a record. Understandably, different readers have different interests, so that the development of labeling practices involves multiple ontologies. Cultural products have to be many things at once—manufactured goods, freight, inventory, consumable, collectable—and each level may require its own labeling and identification of what the product is, while still remaining vested within larger or competing elements of identification. Different products are differently vested. The various readers of film labels remain quite separate compared to the readers of record labels: film producers and exhibitors keep track of things in their lighted offices; audiences keep track in darkened theaters. The former usually read paper; the latter usually read "titles," or labels, on film after they read the publicity poster and the marquee. One way to glimpse the relations of different readers is to examine labeling as a form of substantiation, as a site where the dichotomy between internal and external labels breaks down, or where labels become confused with what it is that they label. The production values of phonograph records and films involved a number of versions of such subtantiation, one measure of which was the simultaneity of labeling and production. One of Edison's laboratory experimenters, for instance, wrote in his notebook:

After close examination of the [phonograph] record handed me by Mr. E I concluded that the article in that form could not be commercially practical, particularly because it did not carry its title[. . . . A] blank must be made of such form that one end is reduced to take the impression of the title at the same moulding as the record itself.[28]

Recording and labeling the phonograph record had to be part of the same single act of manufacture, according to this experimenter. In contemporary parlance, what he was groping for was something like the right mark-up language.[29]

To the degree that mechanical reproduction serves as a particularly ornate label of live performance, labels again recall the laboratory and the self-evidencing that characterizes the invention of all inscriptive technology. Edison's notebooks inscribed laboratory experience, and so did experimental records and films. One of the earliest existing films is a tiny scene of W. K. L. Dickson, the Edison employee largely responsible for Edison's motion picture technology. Dickson plays the violin while two workmen dance for the camera, spinning in their work clothes, their arms clasped as if for a ballroom event. Early commercial films diverged from this homosocial play only gradually; the masculine laboratory culture continued to project itself, as it were, into the new form with its discernible male gaze.[30] Strong men and boxing matches formed early film subjects. Other subjects offered more intricately gendered acts, like *The Kiss* and *Annabelle Butterfly Dance*. Realism had the several affects of play, of prurience and ribaldry. Not surprisingly, phonographs and motion pictures both possess early histories that figure them doubly as toys and as scientific instruments. Among the ancestors of Edison and Dickson's kinetoscope were parlor toys like the stroboscope and the zootrope, as well as the scientific instruments of Eadweard Muybridge and E. J. Marey, both intent on studying the physiology of animal locomotion. Even the telephone was early dismissed by Western Union as a mere "scientific toy."[31] Similarly, Edison's prototype phonograph (so the story goes) first reproduced the words "Mary had a little lamb" to the ears of his astonished laboratory workers. Yet the puzzled U.S. Patent Office classed his invention as a measuring device, a classification Edison long belabored: at the Paris Exposition of 1900 Edison's phonographs were categorized as scientific instruments, while his competitors' machines were arrayed with *musical* instruments. For a time in the 1890s Edison's new language machine pursued a double life, part stenographic tool, part talking doll. In the same years that he helped produce a talking doll (gendered female), who spoke "Mary Had a Little Lamb" and other nursery rhymes, Edison was devoted to marketing the phonograph as the "ideal amanuensis" for businessmen, an automatic stenographer (gendered male). That the prehistories of films and phonographs both resolve mechanical playthings and technical instruments indicates the extent to which both emerged from the laboratory within a performative mystique. They were both cunning objects and privileged subjects, uniquely combining what publicists noted at the time as "novelty" with the lived experience of sound and motion. In this they extended their laboratory roles as the subjects and the instruments of laboratory practice. Early records and films were potently about records and films, in addition to possessing their respective aural or visual performative subjects, dancing workmen or dancing Annabelle.

The musical or amusement phonograph, in particular, entered the marketplace amid residual confusion, doubly figured as both subject and object, ends and means. The most persistent rhetorical feature of this confusion was personification. When Edison had triumphantly demonstrated his first phonograph at the offices of *Scientific American*, the record he played saluted its listeners directly. They reported it "inquired as to our health, asked how we liked the phonograph, informed us that *it* was very well, and bid us a cordial goodnight." Edison's and Bellamy's talking clocks merely continued the trend. While Edison, recalling his early intentions for the phonograph, testified in a contract dispute of 1895 that he had misapprehended the amusement function of the device:

Our views at the time were that the phonograph would [only] be used for amusement purposes in connection with figures, either pictoral or tangible, and would furnish the words or music, or both, which would properly accompany the figures.[32]

By "figures, either pictoral or tangible" Edison meant either pictures of human beings, presented on film or otherwise, or mechanical, three-dimensional figures that actually possessed the shape of human beings. In both cases, his recollected assumption that the amusement phonograph would partake of human form can be seen as the lingering effect of necessary laboratory confusions between the ends and means of technical inquiry. The phonograph displaced human functions (writing, speaking, singing, reading), yet the lexical and perceptual character of the machine's functions remained obscure. Was it right to say that the phonograph spoke and sang? Or were such claims the (anthropomorphic) metaphors of convenience? The answers were yes, and yes. The doubleness died hard. Automatons gave the appearance, at least, of resolving object and subject. They could incorporate the phonograph, embodying and bodying forth its qualities as a machine for talking and as a talking machine. Edison worked on both kinds of figures for phonographs, what he called "pictoral" and "tangible," and many other inventors did too. One of the earliest "tangible" figures involving the phonograph was a fiction: in 1886 the Comte de Villiers de l'Isle Adam published his *L'Eve Future*, a novel in which the character "Edison" builds a woman whose lungs are made of phonographs. (In a feeble act of poetic reciprocity, Edison donated twenty-five dollars toward the erection of a statue honoring Villiers in 1910.)[33]

Only slightly less fanciful were the automatons invented by Francis and James Criswell in 1891 and by George Wilbur Spencer and Alvah Lynde a decade or so later. The Criswells' "continuously and automatically repeating phonograph" was pictured in their patent drawings in the shape of a giant raven. Spencer and Lynde patented a "Speaking Figure," a mechanical human torso containing a phonograph, and having a head with movable eyes and mouth.[34] The Criswells explained that their phonographic raven—reminiscent of Edgar Poe's repetitive bird—was designed to repeat an advertisement for a patent medicine to cure corns. Such figures had their antecedents in the elaborate automata of Jacques Vaucanson and the makers of diversion for the European courts. They in turn proved the progenitors of what Robert Venturi calls the "ducks" of modern life, the overly designed elements of public commercial space.

Nor was the residual conflation of object and subject resident only in the outward shapes and speaking functions of such figures.

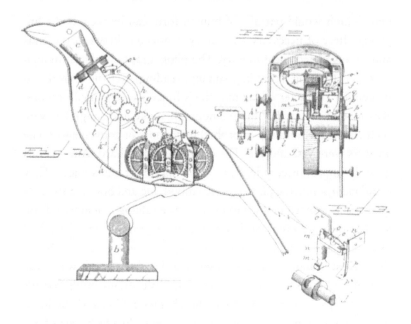

FIGURE 14. Distant relatives. *Above,* Criswell's patent no. 470,477. *Below,* a related duck. (Photo by the author)

The representational nature of recordings for the amusement phonograph retained some of the same problematic, as evidenced at least by ongoing questions of copyright and authorship introduced in Chapter 3. In one law suit brought by Edison interests in Britain, a man was alleged to have duplicated Edison records, going so far as to alter the internal labels recorded on the records. Defendant Percy Henry Levy changed one recorded announcement, "Sung by Victoria Monks, Edison Record," to "Sung by Victoria Monks, Imperial Record," his own brand. The court found Levy in violation of the Merchandise Mart Act, reportedly "for applying a false trade description to his goods, i.e. 'Sung by Victoria Monks,' when, as a matter of fact, it was a copy of the Edison record that was sung by Victoria Monks."[35] Such a case could not have come up in the United States, where duplicating records continued to be a matter of patent law (and under strenuous litigation in *Victor Talking Machine v. Leeds and Caitlin*), but it underscores just how tenuous and arbitrary the boundaries were between performance and reproduction, reproduction and duplication, and correspondingly true and false labeling. Edison records were properly "sung by" the music hall artist because the Edison interests had bothered to find her, hire her, sit her down, and record her voice. Percy's Imperial Records were not "sung by" anyone; they were merely duplicates of the "true" recordings. The internal label, in this case, served as Edison's executives hoped it might, as a form of linguistic checkpoint, a canned shibboleth, admitting listeners to an experience of mimicked performance while barring the way to pirates.

Like the American judge in the *Leeds and Caitlin* case, the magistrate who decided *National Phonograph v. Levy* may not have realized that recording studios then worked by producing "live" master records made by recording artists and that they too duplicated those "sung" master records as many times as they needed. If he did, then his decision located the legitimacy of Victoria Monk's recorded singing in the managerial practices of record production. If he did not, then his distinction between sung and unsung records involved a naive characterization of what the entertainment economy entailed, supposing an overly simplistic commerce, a romantic relation, between recording artist and consumer. In either case what was being articulated was the indeterminate distance between object and subject, between a performing machine and a performing human body, which would be a linguistic and bureaucratic relation to some degree. Such a relation contributed decisively to the construction of new, representational products qua products, as commodities and as inscriptive cultural forms. Not only did mechanical reproduction make entertainment less of a bodily matter, eventually wreaking its particular havoc on the vaudeville houses and music halls, but also recordings mystified the body further. Victoria Monks is lost amid the production practices of the National Phonograph Company, Ltd., and, it seems, in the romantic notions of the magistrate.

Victoria Monks and other performers may even have lost themselves in the production practices of the new mechanical amusements. Different aspects of record and film production remained carefully guarded trade secrets, which served to alienate performers from their own inscribed performances by mystifying the inscription technique. Describing recording practices as late as 1922, one Edison employee ventured, "Few outsiders are permitted to see even the making of a record—certainly no one connected with a rival company." The singer stood before the mouth of a big horn, which protruded from behind a curtain or a partition of some kind; "Even he does not see the actual recording equipment" (Meadowcraft, 698).[36] It was an inversion of the theatrical proscenium, the performer in front and the (mechanical) auditor behind the drape. And film production involved secrets no less carefully kept. Because camera patents became the objects of such litigious controversy, the inner, mechanical workings of cameras became matters of confidence and conspiracy. Producers went to elaborate lengths to protect their cameras from prying eyes and exerted themselves equally in attempts to see inside their competitors' potentially infringing cameras. One federal court judge acknowledged the "tribulations in obtaining the necessary proof" that the camera litigation involved.[37]

Machines and inventors performed in this context, not just the performers who played in front of the camera.

The performing phonograph and the performing inventor were decisively conflated at least twice in these instances of Edison's achievement: first in the talking doll; second in the courtroom, where Edison and his lawyers performed repeatedly. The Edison Phonograph Toy Manufacturing Company was incorporated in 1887, and Edison was contracted to manufacture the dolls. The talking dolls that reached the market in 1890 were complex articles both mechanically and metaphorically. They had human hair, movable leg and arm joints, and contained tiny spring-motor phonographs that played nursery rhymes. Some of them repeated the phrase "Mary had a little lamb" over and over again. These dolls were figured in opposition to their businesslike twins, the stenographic "ideal amanuensis" phonographs. As such they evoked trenchant dualism of identity ranging from child/adult and female/ male, to nursery/office and lyric/letter. Moreover, because Edison's own first words on his first phonograph were widely known to have been the same "Mary had a little lamb," the dolls seem to have been designed to reenact their own invention. They placed that act of invention squarely within the bourgeois domestic sphere of children and women, where they could play upon multiple narratives of generation. The talking doll was simultaneously about the ontogeny and phylogeny of the phonograph, the way the phonograph had first been invented, and the way phonographs continued to be constructed mechanically and construed socially. It was a child's machine about the childhood of the phonograph, speaking within a nineteenth-century context of artificial women's voices that both challenged and enacted the maternal.[38] Children replayed the act of invention again and again as they themselves were inducted into aspects of adult behavior: handling technology is tricky, and the talking dolls needed particular care. In fact, the whole adventure proved a dismal failure. The dolls' machinery broke down, and the nursery rhymes quickly wore off their records. Soon enough, most of the dolls were returned to the manu-

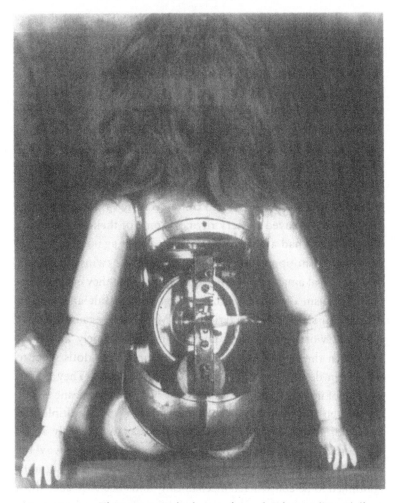

FIGURE 15. The mnemonic body transformed: Edison talking doll (1890). The doll remembered "Mary Had a Little Lamb"; in so doing it kept remembering the invention of the phonograph.

facturer. The invention stopped, and the Edison Phonograph Toy Manufacturing Company folded in 1895, having, perhaps, too precociously mapped the phonograph's path from the laboratory to the parlor (where a few years later cabinet models would become de rigueur), staking out a place in the living environment for amusement machines, marking the territory to be settled by the radio and television in turn. The litigious history of Edison's phonograph is far less a footnote and just as much a record of failed performance. Despite his occasional resounding success in litigation over motion picture technology and his continuous ability to squelch price-cutting wholesalers and retailers of phonographs and phonograph records, Edison lost every major phonograph suit in which he was ever involved, even though he undeniably invented the phonograph. His poor performance has been ascribed to both the evil cunning of his opponents (usually the American Graphophone Company, its president, Edward Easton, and council, Philip Mauro) and the arcanely structured judiciary, which heard patent infringement cases in the federal circuit court presiding in the district of alleged infringement, ensuring that Edison almost never had the home-court advantage.[39] Whatever the origins of his failure, the courtroom tested the inventor, his inventions, and his companies, all in a bundle. In affidavits, depositions, and testimony, Edison and his associates repeatedly reenacted the "working" of his laboratory, inventions, and factory. Laboratory notes, record labels, bills of sale, newspaper advertisements, detective reports, letters patent, and other documents were entered into evidence, reinscribed into the trial record as allies, vouching for the narrative of the complainant or that of the defendant.

The processes and ingredients for making blank record cylinders and the methods of duplicating recorded ones saw particularly heated action. Both litigious subjects became prominent for good reasons; both proved particularly important in the development of the phonograph as a mechanical amusement. Record composition mattered for amusement because harder record surfaces lasted longer and, more importantly, could be played at higher volumes. Tinfoil, wax, and metallic soaps each proved too soft or too brittle to withstand the necessary pressure of a reproducer point that reliably played dance music. For this reason records were one of the very earliest uses of industrial plastics.[40] Methods of record duplication similarly signaled the amusement function and commercial context of the phonograph, since phonograph records used for non-amusement purposes such as office dictation generally did not need to be duplicated. With the commitment to an entertainment function, the developing market encouraged duplication. Musical phonograph cylinders quickly became something like what Susan Stewart calls a "distressed genre" (67).[41] They immediately begged illicit imitation and constituted a whole new form of inscriptive action, shaped by aesthetic and other conscious and unconscious expectations. Some manufacturers sold badly made records imitating the better made ones; others sold mechanically recorded duplicates passed off as "original" or "autograph" recordings. Percy's Imperial Records was just the tip of an iceberg formed of the mass of illicit productions dubiously authorized by manufacturers who did not hold patents. The three major record companies, Edison's National Phonograph, American Graphophone (Columbia), and Victor Talking Machine, had difficulty quashing all the upstarts.

The richness of this imitation and duplication could have deflected attention away from the other generic qualities of phonograph records as simulacra, most centrally their mimicry of live performance. But rather than deflecting attention away from simulation, the free-market economy of imitations and duplicates more likely telescoped the duplicative and mimic qualities of illicit recordings into mechanical reproduction as a whole. In this way, the distress of the genre became tangled with its avowed function and its status as a new inscriptive form. Records remained evidence of themselves, even after leaving the laboratory, but the distinction between mechanical reproduction and duplication proved much less important to consumers than it proved to record companies, composers, and, ultimately, the federal courts.

By 1915 the generic distress of phonographic recordings had so merged with its generic definition and was so accomplished, that Edison's phonograph company manipulated it as a sales ploy, albeit with dubious success. The company staged elaborate recitals across the United States and Canada, in which live performers appeared on stage beside Edison phonographs playing recordings of their own voices. The "tone tests," as these promotions were called, always included a moment at which, with the lights dimmed, the audience heard only one voice. Contemporary newspapers reported that people were frequently fooled and their pages were larded with advertisements in an all-out public relations campaign. The lights would come up, and the voice the audience thought was human proved to be mechanical. The phenomenon possesses broad implications for the cultural history of musical fidelity and perception,[42] but it also, like Edison's talking doll, returned the phonograph to its origins, to the *locus mirabilus* of the talking machine, where audiences gasp upon hearing a machine sing. The tone tests were a throwback, but they punctuate a new relation between the body and evidence, between the self and self-registering instruments.

~

Though the parallels between the professionalization of science in the eighteenth century and the emergence of an entertainment economy in the nineteenth can quickly be overdrawn, the comparison is nevertheless a telling one. Just as scientists came to rely upon the readings of instruments rather than the vagaries of bodily experience, modern entertainment renegotiated the authority of the body in its embrace of mechanical reproductions such as photographs, phonograph records, and films. Edison's laboratory can be seen as one membrane through which the two relate, his scientific and technical work connecting to what consumer culture conceives as play and pleasure. Because Edison worked on so many technologies of representation—communicative media like the telegraph and telephone, but especially inscriptive devices like the phonograph— the discursive practices of his laboratory and factories are particularly illuminating. Just as changes in scientific practice can be gauged in the scientific article's generic definition, rhetorical composition, and patterns of circulation, so too can changes in the culture of entertainment be gauged in the changing labels entered in and onto new inscriptive products. Developing practices of labeling for phonograph records show the rhetorical negotiation of what records were perceived to be, explicitly characterized by their producers within a context of ongoing market exchange. The hard-won parlance of the *record label* dies hard: the term itself is a rootless synecdoche today, standing for a corporate entity producing compact disks, not records.

During the period from 1895 to 1909 particularly, phonograph records and motion picture films occupied a confused place as both subjects and objects of attention. Their inherent slipperiness as articles of manufacture and sale was visible in the profusion of labels they received. It was further indicated in what became their generic distress, as records and films were widely pirated and falsely advertised. This slipperiness was also obvious in their dual origins as parlor toys and scientific instruments. Phonograph records emerged from Edison's laboratory amid residual confusion concerning their status as evidence, vouching doubly for themselves as incompletely legible inscriptions and for the human performers they reproduced. Their use in dictaphones perpetuated their status as instruments, though market commitment to their use as amusements was relatively swift and nearly total. While early-modern science had seemingly purged itself of the human body, the entertainment phonograph mystified the body in talking dolls and tone tests and, more particularly, in the production values of the record companies. Records placed the performing artist and the listener, Victoria Monks and the magistrate, at an indeterminate distance from one another. The distance was partially articulated by internal and external labels, by new requirements for handling and demonstration, by record quality, and by other features of the newly figured commercial status of entertainment, not the least of which were the loudness of dance music and the magistrate's romantic misapprehensions.

Technology transfer from the drawing board to the marketplace involves complex rhetorical and other acts that turn the combined subject and object, the ends and the means of laboratory inquiry, into a product. As a product, the invention is supposed to become the object of consumer desire. But with cultural forms like phonographs, phonograph records, and films, the toggle between laboratory and marketplace seems to remain notably imperfect. As media, cultural forms do not lose their complicated status as instruments, harbingers of objectivity, of a rhetorical authority. But neither do they continue unabated the realist or objective purpose of laboratory instruments. They proffer additional subjectivities, like the production values and performers they inscribe and promote, with all of their incumbent indeterminacy, cloudiness, and noise. The "noises of speech," as scientists noted right away about the phonograph, get lost in mechanical reproductions. "Noise" is Jacques Attali's term for the political economy of music. It is similarly William Paulson's term for encoded culture, for "anything that gets mixed up with messages as they are sent" (ix). Its very mixed-up quality of inarticulation makes noise difficult to identify and explain, particularly in the raucous curiosity of considering literature or music as "messages," apart from or in addition to considering them the aesthetic subjects of cultural analysis. New media clearly involve new kinds of noise, as indicated by turn-of-the-century changes to record and film labels. There are certainly other examples. The next chapter explores a group of associated taps and mummers comparable to, yet distinct from, the noise of mechanical amusement. New business machines such as typewriters made inscription sound differently than it had before, both in the literal clatter of keyboards and in the background hum of accommodation, the regularized features of business typing and the context of the typing subject.

5

Automatic Writing

[The typist] smoothes her hair with automatic hand
And puts a record on the gramophone.
  —T. S. Eliot, *The Wasteland*

The concept of the phonograph came to Edison when he was a consultant to Western Union, working to circumvent Alexander Graham Bell's patents and improve the telephone. One of Edison's laboratory employees, Charles Batchelor, recalled that, at the time, "Mary had a little lamb" was one of the "favorite stereotyped sentences used in experimenting on the telephone." The word Batchelor chose to use—stereotyped—refers to a printing process called *stereotyping*, used widely by book publishers. His phraseology, "stereotyped sentences," both demonstrates the early metaphorical valence of the word stereotype and tacitly underscores the problem with telephony. Edison, Western Union, and an industry familiar with printing telegraphs and tickertape thought the telephone too ephemeral because it lacked an inscribed record. Edison and his laboratory staff had little time for anything but telephony in the several months between the concept and prototype of the phonograph, though they did work frequently with one structural component that proved vital to the new instrument, a rotating cylinder. Rotating cylinders were part of

Edison's repertoire as an inventor; he returned to that shape again and again for different purposes—in telegraphs, autographic printing devices, and telephones, and then in the "mandrel" of his phonograph, his first attempts at motion pictures, and subsequent ore-crushing machinery. The rotating cylinder has been described as part of Edison's "style"; it was his version of the baroque writer Thomas Browne's quincunx, a shape that looms everywhere once the search is on.[1] While the functional diversity of this shape is some indication of the varied textuality Edison and his staff experienced in experimental telegraphs, printers, and the like, its comparable public apprehension remained more limited to phonograph mandrels and typewriter platens.

This chapter shifts the axis of inscription, from the spirals winding around a cylindrical phonograph record to the alphabetic characters, spaces, and nonalphabetic signs that move across the typed page. Like shorthand alphabets and phonographs, typewriters were appropriated within textual practices, which they also had a share in transforming. Phonography and phonographs converted aural experience into inscribed evidence, the former by representing the sounds of speech on paper, the latter by reproducing the same sounds in the grooves of a record. Typewriters intervened more directly into the experiences of writing itself in ways that further interrogated categories of orality, aurality, and textuality. By staying the course as a textual device, typewriters formed an important compliment to phonographs, which ultimately diverged dramatically to amusement purposes. The amusement phonograph retained some of the power of text, enrolling authors as producers, writing as hieroglyphics, and machines as readers. Typewriters, however, seemed more certain instruments of textual production, making up in the clarity of their typeface presentation what they lacked as a means of wholly objective representation or reproduction. In place of objectivity, which for phonographs was the sounding "just like" that records did, typewriters helped negotiate new forms of human subjectivity that supplemented the experiences of chirographic and printed authoring. Some adaptive sources of this subjectivity were predicable, for instance, the intermediate position of the shorthand reporter as a scribal technician. Still others proved surprising. In this chapter, I will demonstrate how the connections between psychology, spiritualism, and typing in the 1890s find resolution within the term *automatic writing*. This term was applied to writing done by partially conscious, distracted, hypnotized subjects as well as to writing done "mediumistically." The term was also widely applied to typing.

Psychology was an emerging "human" science at the end of the nineteenth century. American advocates and practitioners of this new science were anxious about its disciplinary, scientific status even as they contested its subjects, methods, and nomenclature. The language of psychology was hybrid, combining "old philosophical patois" and physiological terms within a self-consciously scientific rhetoric of experimental results. One source of psychology's rhetorical authority was its avowed distance from the related field of psychical research, though some, like William James, continued in the interdiscipline. The scientific status of psychology rose as the comparable status of psychical research sank, tarnished by its association with parlor seances and popularly reported spiritualist "phenomena."[2] The *Psychological Review* became, after its founding in 1894, an authoritative organ for the new psychology in America. Early contributors included James, Hugo Münsterberg, and other notables in the field. The new journal contained limited advertising space, usually just a page or two per issue. In a 1895 issue, the Columbia Typewriter Manufacturing Co. advertised its Bar-Lock machine:

Writes every letter in sight of operator. Does most of the work, in writing *automatically*.[3]

This coincidence of automatic writing as both psychophenomenon and as typing points doubly to the openness of the word *automatic* during the 1890s. It also highlights attitudes regarding writing, or, more broadly, cultural assumptions regarding the act of inscription and the relation that inscribing bears to authorial agency and textual evidence. In distinct ways, the development of the typewriter and the interests of the new psychology involved questions of authority and witnessing, or visuality, and the dynamics of textual production. Similar issues emerged in typewriter design, the cultural construction of both typing as an activity and typescripts as writings, and in the disciplinary bounds and methods of the new science, with its location of professional practitioners over and against amateur practitioners and the human subjects of practice. The position and composure of the authorial subject was a point of anxious negotiation in both instances. The connections between authorship and writing became attenuated and obscure when the latter became newly "automatic." The unimpeachability of writing as evidence was itself rearticulated in the varying circumstances of automaticity, which were at once material, bureaucratic, and anatomical. A whole new *discourse network* emerged, to borrow Friedrich Kittler's term, a whole new system of making written sense.[4]

What was "automatic" in the writing studied by psychologists and psychical researchers, how was it evidence, and why was it important to them? How was typing construed as "automatic," and was it evidence of anything at all? This chapter addresses the first of these questions in a brief reading of an article by Gertrude Stein, written for the *Psychological Review* when she was a student of James and Münsterberg at Radcliffe College. The chapter then addresses the second of these questions in some general observations on the technical and commercial history of the typewriter as a business machine. In addition to their mutual bearing on the authority of inscriptions, the two paths of inquiry inform each other with regard to an implied subject—gender. Psychology (and certainly the emergent psychoanalysis) figures women as subjects and has historically implicated gender in its construction of normality and pathology. Psychical research seems to have been chronically divided between, in James's words, "the scientific-academic mind" of professional inquiry, and "the feminine-mystical mind" of amateur dalliance and Theosophy (Murphy and Ballou, 27).[5] For its part, secretarial typing forms the classic case of the sexual division of labor.[6] Office work once done wholly by men was later done wholly by women, the typewriter reportedly having played a central role in this dramatic reversal.[7] Just as the "reading machine"-phonograph described in the U.S. Copyright Act of 1909 provoked (not wholly unpleasant or unprofitable) anxieties about the racial identity and the mimetic remoteness of performers, mystified in the mass market, the new "writing machine"-typewriter similarly involved some defining anxieties about intention, gender, and other conditions of authorship.

The typewriter and phonograph are almost exact contemporaries. The phonograph, invented in 1877, reached the market as a dictating machine and embarked upon its career as mechanical amusement in the early 1890s. The first typewriter reached the market in 1874, impressed observers at the Centennial Exposition in Philadelphia, and seems to have arrived by 1894 at the general form and purpose it would keep until challenged by the personal computer almost a century later. The typewriter and phonograph were partners for a time: the local phonograph companies leased their dictation machines to sit on desks beside typewriters; the former to record and the latter to transcribe. Both arose from the same climate of representation, the same urges to inscribe lived experience, and similar assumptions about language and machines. Both seemed at the time to embody some progress beyond shorthand and other, similarly regular(ized) literacy practices, like penmanship, calligraphy, even telegraphy and musical notation. Yet the features of that progress remained obscure, clouded by emerging economic and bureaucratic structures, by new tastes, and new varieties of powerfully technological experience. Differences between phonographs and typewriters, like those more broadly between popular culture and office work, cannot hide their complementarity as registers of the changing nature of late-century subjectivity. Mechanical amusements such as phonographs and motion pictures harbored resilient objective qualities while at the same time mystifying the identity and remove of the performer and the aura of origination, the inscriptive process and the represented experience. Typewriters, meanwhile, more directly challenged the author as agent by offering a newly mechanized, newly gendered, and self-consciously "managed" imposition between the mind and the page. Typing obscured writing while it provoked a new awareness of manual, visual, and aural habits at the heart of vernacular literacy practices.

Before going any further, however, it is necessary to recoup some of the less familiar connotations of the word *automatic* in the 1890s. A relatively modern word, coined in the eighteenth century and given general use in the nineteenth century, it derives from the older *automaton*, also has lingering connotations of resolving the organic and mechanical—of human forms and functions built into machinery and of mechanical responses by human beings. "Automatic" was not a simple synonym for "mechanical," nor a simple antonym for "by hand." One spin-off Edison product of the 1890s and 1900s makes the point directly: the Edison or the Bates Automatic Hand Numbering Machine, a hand-operated stamp for inscribing sequential numbers on bills, invoices, and the like. Like the explicitly "manual" typewriter of the 1920s and later, the "hand" numbering machine was no less emphatically "automatic" because it used a human source of kinetic energy. The numbering machines were "automatic" in that certain features of their operation did not require human attention; the numbers advanced consecutively and reinked as the stamps were used. These were both labor-saving features and a matter of "accuracy," as the promotional literature explained. The largest users of the machines were "insurance companies, trusts, banks, railroads, and government departments," bureaucracies necessarily rich in paperwork and where this form of inscription possessed a vested institutional authority according to its uses and its rhetoric of numerical certitude. By implication, automaticity involved more than just the replacement of concerted human attention. Automatic functions fractured and displaced attention within bureaucratic and technical systems. Users of the hand-numbering machine did not have to think about advancing and reinking the number stamp; those functions had already been thought of by the men who hired the users and purchased the numbering machines. Work rules and the process of machinery combined to offer businesses a form of "mechanical objectivity," a presumed freedom from human subjectivity and, consequently, from error.[8]

Like concerted attention, however, subjectivity could hardly disappear, and the American desire for more and more automaticity took some ridiculous turns. The first distributors of Edison's phonograph, for instance, worried that their machines required too much attention of users. Attention in this case meant skill, patience, intelligence, and the pure motives of a worker who wanted to maintain and operate the new, temperamental device as a dictaphone. The men at the local phonograph companies convention did not see requirements that thoroughly, however; they simply called for "a machine which requires as little work as a telephone or sewing machine or reaper or any other automatic machine" (*Proceedings*, 58). Perhaps none of them had experience operating reapers or sewing machines, which were "automatic" perhaps, but not robotic, as they seemed to think. Sewing and reaping are both hard work. The facile confusion of work and attention proved endemic to celebrations of automaticity and the mechanization of labor. *Automatic machine* became a confusing generic term; exploring its semantic range helps historicize *cyborg*, a weighty term in today's critical theories of culture.[9]

One particularly pertinent semantic field for "automatic" was the nineteenth-century language of telegraphy. Like the railroad, the telegraph was a dominant technology of mid-nineteenth-century America—in the sense that the economic organization of the century depended upon it with ever-increasing commitment. Unlike the railroad, however, the telegraph was a form of infrastructure that remained at one remove from the daily lives of most people. Popular experience of the railroad as a feature of nineteenth-century culture and environment has been amply documented. The telegraph resisted the same familiarity. Whereas the railroad possessed its own professional "code" in the form of technical and corporate practices, the telegraph seems to have possessed a code that insulated insiders from outsiders to a greater degree.[10] The dots and dashes of Morse code are only the most obvious ingredient of this professional discourse, impenetrable to the uninitiated, which extended to features of corporate practice, wire services, electromagnetic and technical process.

It is unclear, therefore, to what degree and with what specificity the notion of "automatic telegraphy" entered the popular imagination as it consumed the attention of Edison, his associates, and counterparts in telegraph workshops across the United States and Britain during the early and mid-1870s. Technical experts and telegraph companies, both upstarts and established concerns, set their sights on a group of related innovations—among them automatic telegraphy, printing telegraphy, and duplex/diplex telegraphy—working within what have been characterized as the first programmatic research and development initiatives. Automatic telegraphy involved using a perforated strip of paper, with holes representing dots and dashes, to transmit a message that was then recorded on another strip of paper as it was received. Printing telegraphy used electrochemical or other means to print the received dots and dashes. In its most keenly sought form, it translated these dots and dashes into Roman letters. Duplex/diplex telegraphy sent two messages across a wire at the same time. Edison worked feverishly on all three, and in 1871, he combined automatic with printing telegraphy in some work on what one of his backers, Daniel H. Craig, dubbed the "'*Automatic Writer*' (How'll *that* do for a name?)" Craig's baptism of Edison's work-in-progress is immediately interrogated by Craig's self-conscious question, suggesting either that the words "automatic" and "writer" had never before been dreamt together in Craig's philosophy, or that he was drawing ironically on the language of spiritualism.[11] The impetus for such innovations in telegraphy was twofold. First, automatic telegraphy was fast; running a strip of perforated paper through a transmitter sent messages at far higher speeds than tap-tapping on the old telegraph "key" did. Second, automatic telegraphy was cheap; perforating, transmitting, and receiving machines could be handled by low-paid "girls" instead of by more highly paid male "first-class operators," as the distinction was often drawn. The word "automatic" in this context denoted an increased efficiency because it was mechanical, and a decreased skill level for operation that was pointedly feminized. The concerted attention that skilled operators directed at their instruments was displaced onto the concerted attention that employers could now pay to the "girls" and to other technical and nontechnical features of their systems.

Both automatic and printing telegraphs resulted in portable inscriptions; the incoming message was in the form of an inscribed record, not an ephemeral tap-tapping sound or visible oscillation that had to be recorded as it arrived by a skilled operator with a pen and paper. These inscriptions could be distributed and processed by large groups of people as they arrived. Daniel Craig found that automatic telegraphy "was tumbling messages into the office" at 1,000 words per minute, far too fast for even a handful of people to copy efficiently with what he now called the "hand-pen." The typewriter was therefore adopted to process the glut of messages, even before the machine was put on the market by the Remington arms and sewing machine manufacturer in 1874. Craig seems to have been using a version of the Sholes machine (eventually the Remington) in 1871, and Edison finished his own model of what he called a "Mechanical Printer" in May of the same year, after having seen a Sholes prototype in 1870. He continued to work on the machine for a number of years and received his first patent in the area of "Type-Writing Machines" at the end of 1872, for a machine vastly different in its mechanics than Sholes's but identical in its result.[12] Like the elaborate stenciling pens and other textual duplication devices that also absorbed his attention, Edison recognized the typewriter as one component in a successfully "automatic" system of business communication. This was the kind of systematic thinking he would later bring to electric light and power distribution, with its manifold components. Like other parts of the automatic telegraph system, the typewriter had the advantage of being fast *and* efficient, much more so than the "hand-pen": it "PILE[D] AN AWFUL STACK OF WORDS ON ONE PAGE," as Mark Twain put it on his early Remington, which had no lowercase type. And though it required a very specific skill acquisition to use effectively, the typewriter resembled other components of automatic telegraph systems as a site for feminized labor.

These connotations of "automatic" likely informed the "automatic writing" of the later century, though spiritualists and psychical researchers admittedly wrote of "mental telegraphy," rather than specifically "automatic" telegraphy. As a dominant technology of the period, telegraphy did possess a great deal of metaphorical weight, despite (or perhaps because of) the fact that relatively few people had direct experience with or specific, professional knowledge of the telegraph. This symbolic weight is evidenced lexically in that telephony was an early form of "acoustic telegraphy," and the telephone was briefly called the "speaking" telegraph. The promise of what were termed *aerial telegraphy* (i.e., radio) and *mental telegraphy* (extrasensory communication) did not seem too far fetched to Mark Twain, Edison, or a vast number of others. Edison joined the Theosophical Society in 1877, and as late as 1910 explored phenomena of mental telegraphy performed by the spiritualist, "Professor" Burt Reese. Twain corresponded with the Society for Psychical Research in 1884, affirming his belief in mental telegraphy, which he wrote about on several occasions.[13] Spiritualists and psychical researchers stuck to telegraphy as a metaphorical model for "spirit" communication. They cocked their ears for the "sort of desultory, telegraphic tattoo" of spirits knocking and rapping at seances; this according to William Dean Howells's spiritualist novel of 1880, *The Undiscovered Country* (26). In a typical, nonscientific account of automatic writing, one of Edison's unbidden correspondents made the connection between that writing and mental telegraphy explicit:

I discovered last fall that by holding my pencil in a certain position, it would write things without any conscious assistance on my part, and after many experiments concluded that *electricity was the medium and my hand and pen the telegraphic instrument as it were, of the invisible operators.* (Simson 1911; emphasis in original).

The writer of this idea letter, a self-described Quaker from Emporia, Kansas, who manufactured and sold eye ointment, doubles the function of "medium." Both he and "electricity" serve as means of communication. His personal revelation, made semipublic in his correspondence with Edison, is an authorial vignette that collapses on itself as its author's categories of visuality and instrumentality contradict themselves: a purveyor of eye ointment who discovers invisible operators, a letter writer, pencil holder, *and* human telegraph key, Mr. Simson leaves strenuously unacknowledged any coincidence or disjuncture between the inner light of Quakerism and the automatic telegraphy of spiritualism.

Because it was less interested than spiritualists in communication, the new science of psychology was also less interested in telegraphy. Yet it remained interested in things automatic because of its focus on anatomizing human attention and consciousness. In 1896 a young Gertrude Stein coauthored with a Harvard graduate student named Louis M. Solomons an article on automatic writing for the third volume of the *Psychological Review*. As Stein put it in her subsequent solo article, she and Solomons were their own instruments of study; "The only subjects we had were ourselves." In 1898 Stein sought to improve on their findings by "using a large number of subjects." The resulting article, subtitled "A Study of Character in Its Relation to Attention," details Stein's experiments with automatic writing observed in "normal individuals," all students at Harvard or Radcliffe and mostly what she called "self-repressed" New Englanders. Stein's experiments involved the "cultivation" of automatic writing. Her subjects were made to devote their attention wholly to reading or conversing while she guided their hands in some rhythmic motions outside their fields of vision. What she wanted to see was whether, when she stopped guiding them, or if she pretended to guide them, the subjects would continue with any kind of automatic writing. Her interest was less in what the subjects automatically wrote than in whether and what kind of automatic motions they developed, and what, if anything, that might imply about the "characters" of these "normal" college students, including their gender differences. Others who studied automatic writing laid more emphasis on what was automatically written. This was particularly true in parapsychology, in psychical research, and in Theosophy, where extrasensory perception and spirit communication seemed evident in the contents and kinds of automatic writings. Madame H. P. Blavatsky, celebrated medium of the 1870s and founder of Theosophy, is said to have experienced six different kinds of automatic writing during the generation of her orientalist Theosophical doctrine, among them "writing by dictation," "writing by directive clairvoyance," and "writing by psychometry."[14] The types varied in their substance, presumed process and creative origins, and even in their handwriting; all six were differently automatic.

A similar impulse toward classification characterized professional psychical research and the science of psychology. Automatic writing was only one in a recognized hierarchy of phenomena. In 1904 James distinguished rudimentary "motor-automatisms" from "various alphabet-using forms of amateur mediumship," like the Ouija board. These in turn were distinct from "graphic" automatisms, of which automatic writing "is the most popularly known example" and quite common. In his standard textbook *Principles of Psychology* (1890) he had drawn similar hierarchies, going so far as to identify practiced "playing on musical instruments" as belonging to "the relatively lower phases of possession" (Murphy and Ballou, 48). Both psychical research and psychology acknowledged a difference between automatic writing done by self-aware subjects and hypnotized ones. Such classifications were theoretical as well as practical: Stein found that the cultivated automatisms of her subjects divided them into two groups. Types of response corresponded to types of people. Type One consisted "mostly of girls who are found naturally in literature courses and men who are going in for law." She described them as "nervous, high-strung, very imaginative," possessing "the capacity to be easily roused and intensely interested." Type Two consisted of individuals "either large, healthy, rather heavy and lacking in vigor," or in "what we call anemic and phlegmatic" types. They possessed very little "power of concentrated attention" (Stein, 297).

At stake in all this classification was a definition of consciousness, an articulation of mental process and personality that was decidedly unpolar according to the now familiar consciousness/unconsciousness model, though it was polar in other respects. It questioned the resemblances between motor and mental habits, suggested kinds of symbolic activity, and interrogated aspects of attention.[15] Stein's psychological experiments, in particular, divided her Type One from Type Two; she found less difference between the sexes than between the two types, a convenient finding, since her experiments could then confirm the ones she and Solomons had made on themselves. She and Solomons had both experienced a lot of trouble disassociating their roles as scientists and subjects. As hard as they tried, they could not stop paying attention; Stein admits, "It was our inability to take our minds off the experiment that interfered" (305). They got plenty of writing (including, by extension, their scientific article), but little of it was successfully "automatic." In the later experiments Stein concluded that Type One subjects responded more like she and Solomons had when they failed, and Type Two subjects responded more like she and Solomons had on the occasions when they succeeded in writing automatically. It seemed to Stein that Type Two subjects were "much nearer the true hysterique, where powers of attention, or rather lack of power of attention, induced an extreme suggestibility and a great tendency to automatic movement" (305). Her polar classification of types was validated to the extent that it corresponded with psychology's established dichotomy between normality and hysteria.

That Stein herself ended up with the more normal type of character, self-possessed, vigorous, with more power of attention, could have been a matter of fleeting satisfaction for her readers, who were caught in a paradoxical relation to her evidence. In self-defining normality, Stein was arguing partly on the basis of experiments that failed; she, Solomons, and the Type One subjects did not write very automatically at all in this study of "normal automatisms." She was also compromising the assumed objectivity of her work, which possesses a dubious "falsifiability" in Karl Popper's sense. Stated in the extreme, her conclusions resemble the Cretan paradox: her results could be trusted if she were normal, or her results could not be trusted if she were not normal. Such was the price of being both psychologist and subject, scientist and instrument, at the same time. Readers of her article seem to have accepted her findings in part because they accepted her normality. In this they relied upon rhetorical, institutional, and other features of psychological discourse. Her article fit the genre. Her profile—her Radcliffe affiliation and her mentors—fit the profession, and her normative division and descriptions of character fit the discipline. (That her later life as a lesbian expatriate writer would not have been considered normal by the same readers is only the sharpest reminder that normality, like the other features of psychological discourse, is a protean construction.)

Professional psychical researchers seem to have sensed themselves infected by a worse case of the same paradox affecting the new psychologists.[16] It would be difficult to affirm paranormal findings by normal subjects and normal investigators with anything like scientific authority, since that authority was based in part on vacating the "para" in paranormal. William James and the psychical researchers reacted to this bind by devoting themselves wholeheartedly to the scientific method. In particular, they focused on rigorous documentation of their work, on the means and methods of reporting. The popular literature on the supernatural was huge, but it had to be discounted, James warned, as "practically worthless for evidential purposes. Facts enough are cited, indeed, but the records of them are so fallible and imperfect" (Murphy and Ballou, 30). James had chided in his first report to the American Society for Psychical Research that professionals should obtain "verbatim stenographic reports" of their experiments with mediums. He wished aloud in the same report that the society could find a control subject, a good medium "at the outset of her or his career." Before engaging in any "miscellaneous work," the medium could be "patiently and thoroughly experimented on, with stenographic reports of trances, and as much attention paid to failures and errors as to successes" (96, 99). The society never fulfilled James's wish, but its members set about collecting evidence, meticulously documenting their work and observations and elevating the standards of their reporting. Stenographic transcripts, the written testimony of witnesses, and paper exhibits formed the raw language of psychical research, assimilated and offered to the scientific community in conventional scientific articles. The legalistic temper of the psychical observers and their inscriptions is telling. Like litigants presenting evidence in the courtroom, the psychical researchers presented evidence before the high court of "science," that is, before the culturally authoritative Laws of Nature. In doing so they relied upon a mechanist model of writing in which the speed, efficiency, and immediacy of descriptive compositions vouched for the accuracy of described observations. Psychical researchers aspired to the inscriptive authority of stenographers and court reporters, who themselves shared so much ambivalence about the degree to which their profession involved being automatic. Researchers sought to normalize paranormal phenomena by representing them within official texts legitimated by the objective intentions and the on-site notations of reliable witnesses, the more professional the better. Inscriptive method validated inscribed content.

Even the popular literature that James discounted, of which there was a great variety, tended implicitly to privilege method in its presentation of fact. In calling for more psychical research from the pages of the monthly, *Arena*, B. F. Underwood wrote at length about the automatic writing performed by his wife. He had observed her "closely for nearly two years" and reported that she wrote automatically in a "normal" state of mind, not in any trance. She writes down a question, "and then waits for an answer, which is written rapidly under the question." These automatically produced answers make her hand tingle and are characterized by spelling, style, and handwriting distinct from her usual. Her body acts as an uninvolved conduit for the writing; that makes it "automatic." Mr. Underwood, by contrast, is not bodily or mentally involved, except as an observer: he assures his readers, "I in no way consciously contribute, directly or indirectly, either to the writing or to the thought expressed in these answers" (Underwood, 76). His attention is to the phenomenon; curiously, his presence seems to be a prerequisite for its occurrence. His article represents the couple as a symbiotic pair—she, the medium, abnormally subjective, and he, the observer, carefully objective. He mindful, she mindless.[17] It is a symbiosis frequent in the literature of psychical research and revealing of its method, an enacted Cartesianism, a mind/body (so frequently his/ hers) dualism that implies the legitimacy of a paranormal phenomenon by touting its rigorously normal, objective description by assuredly impartial witnesses.

The heterogenous *Arena* was "the leading radical magazine of that day," but it also offered a forum for psychical inquiry that possessed neither the sobriety of the *Proceedings of the American Society for Psychical Research* nor the commercial biases of the "badly printed spiritualistic weeklies" available at the time (Garland, 1, 4). A devoted contributor to *Arena* was the young naturalist writer Hamlin Garland, who, like several other authors of his day, came to have an interest in psychical phenomena.[18] Late in life Garland wrote a memoir of his "Forty Years of Psychic Research," which gives a good idea of the sort of popular mediumship available during the period. Garland details his inquiries into "slate writing" and "trumpet mediums." The former involved writing that mysteriously appeared on slates that were fastened together, face to face, and the latter involved voices that mysteriously spoke through metal cones during seances. Both were practiced by numerous mediums, a vast proportion of them women, and were presumed by the mediums and their devotees to be means through which so-called "invisibles" (the dead) communicated with the living. Both were elaborate variations on the standard seance themes of "floating" tables and tapped-out answers "transmitted" and "received" in response to questions. Both seem to have served as intricate, metaphorical responses to seance constituency, which always included skeptics as well as devotees. Slate writing, with its connotations of grammar school pedagogy, was positioned to teach the skeptics and to affirm a tutelary balance of power in which the dead instructed the living. Trumpet mediumship likewise configured its audience by evoking a balance of power. The ear trumpet was an accouterment long associated with the hearing impaired, while the horn of the commercially available phonograph possessed a similar shape and potently resolved the functions of ear and mouth. Trumpet mediumship was positioned to make the deaf hear and the mute speak, much in the same way that the phonograph was early supposed to let the dead talk and the blind read.

Garland, a skeptic, put both phenomena to repeated tests, which he describes in exhaustive detail. Like most seance participants, the evidence he usually asked for was self-knowledge.[19] He challenged slate writers and trumpet mediums to tell him about himself, his family relations, friends who had died or who had sent him to the medium. In this way the medium, the "invisible," and the inquiring participant all shared the position of subject. Seances were about all three of them. Garland of course tried to ward off fraud. One series of experiments with a trumpet medium involved what amounted to a sadistic bondage ritual of tying, taping, and nailing the woman medium to her chair for five or six hours at a time, to be sure that she did not perpetrate any fraud while "the sitters" of the seance sat in the dark making various contact with "the invisibles" (55 and following). Contact was evidenced in trumpet voices and tapping sounds, which came in answer to questions posed by the sitters, as well as in the inexplicable movement of physical objects around the darkened room. The bizarre rigors of Garland's investigative method are obsessively detailed in what he entitled his "plain narrative of fact," in part because Garland wants the "plainness" of his account to vouch for its "facts." Like the tale of Mr. Underwood, the reasoned, literate comportment of this male eyewitness seeks to validate the unreasonableness of what he saw and heard, the possession of a passive, female medium by invisible spirits of the dead.

∼

In different forms the same attention to inscriptive method had long characterized the literature on occult phenomena. William James and the professional psychical researchers only wanted to be newly systematic and bureaucratic. Similar impulses clearly characterized other disciplines and institutional and social activities during the second half of the nineteenth century, evidenced most pertinently in the acceptance and growth of stenography and the eventual success of typewriters. Accuracy in stenographic reporting entailed a verbatim record made wholly on the spot by a specially trained and specifically affiliated reporter. By 1900 a number of specially designed inscriptive devices were used in the courtroom, precursors of the present machine. The result was an essentialist and authoritative transcript that both described and comprised "what happened" and "what was said." In order to work this essentializing magic, court reporters and stenographers negotiated an intermediate position between experience and inscription. In doing so they represented a synthesis between (translated into spiritualist terms) the sitter and the medium. They served as both reliable witnesses and uninvolved conduits, Mr. and Mrs. Underwood, thus disrupting the Cartesian divide between researcher and subject and shifting the nature of inscribed evidence as they articulated the role of scribe-technician. While the mid-nineteenth century had witnessed a great deal of foment over varying systems of shorthand and the nature of authoritative transcripts, the related construction of the scribal technician proved of even greater consequence. As a technician, enrolled within the rule-based practices of literacy and bureaucracy, the reporter and secretary possessed an associative authority. Yet particularly where new technologies joined the scribal technician between author and text, the technician's own position acquired a curious vagueness or liminality. Gender difference marked this liminality in many of the inscriptive and noninscriptive practices of the nineteenth century, including mediumship and typewriting but extending as well to telegraphy and the telephony of so-called "Hello Girls" or telephone-exchange operators. The same indeterminacy came differently marked: it may help explain that, much earlier, the first family of British civil stenography, the Gurneys, were Quakers, or even that the American federalist reporter, Thomas Lloyd, was Catholic. Difference, in whatever form it took, became party to the very generation of authoritative text in the person of the scribal technical. The margins performed at the centers of power. That is one of the points Edison made when he recalled his youthful exploits:

As I thus kept posted & knew from their activity every member of Congress & what committees they were on & all about the topical doings as well as the prices of breadstuffs in all the primary markets, I was in a much better position than most [telegraph] operators to call on my imagination to supply missing words & sentences which [were] frequent in those days with old rotten wires badly insulated[,] especially on stormy nights—on these nights I had to supply in some cases 1/5 of the whole matter—pure guessing but I seldom got caught except once. (Edison, *The Papers of Thomas Edison*, 1:657)

Edison writes himself into a literal gap in some rotten telegraph wires and also the more complexly featured gap between author and transmitted text. Here Edison was in "a much better position" because he kept up with current events, but he is no less marginal for that better position when it comes to the actual political events of his day. The very conditions of his employ put him at the in-between, at once central and marginal. (The one time he remembered getting caught was when he "reported" an event that had been postponed.)

Even more than the young Edison, James O. Clephane seems emblematic of varied inscriptive means, interests, and subjectivities. Clephane was a shorthand reporter, a typewriter promoter, and a capitalist involved in the phonograph and the Linotype. Clephane started his career as a congressional reporter. In that capacity, he was questioned about Pitmanic phonography during the impeachment trail of Andrew Johnson. He also lent his support to Christopher Latham Sholes, pushing work on the early typewriter. In 1876 he was the Washington, D.C., agent for Edison's electric pen, a shortlived textual duplication device. Clephane helped inspire Ottmar Merganthaler to embark on the developmental work that would end with the Merganthaler Linotype machine. He served on the boards of the Linotype Company as well as the American Graphophone Company, and had earlier attended the local phonograph companies' conventions as a backer and executive of the Eastern Pennsylvania Phonograph Company. To a surprising degree he personified the combination, competition, and proliferation of inscriptive means that characterized American culture and bureaucracy from the Civil

War until his death in 1910. True to the nature of newly automatic writing, Clephane proved an illusory, liminal figure to the last. His obituary lists the many important men he had known, as a reporter particularly, with the note that he was "associated through them with many of the most stirring events of the civil war period."[20] Clephane had been neither enzyme nor substrate, author nor reader. He was associated with stirring events, not as an actor or subject but as an inscriptive agent. His liminality certainly had little to do with his personality and much to do with the contradictions of being at once central, in between, and marginal.

Predictably, the role of scribal technician was circumscribed by shared assumptions regarding civil and social authority, the abilities of the human body and mind, and the increasing relevance of technology to matters of writing and reading. There was an underlying conflict over how much intelligence the scribal technician had to supply, when and how much the head and the fingers worked, just how automatic stenographers, telegraphers, and other scribal technicians had to be. Not surprisingly, the presumed precision and self-less operation of machinery provided a metaphor for the newly systematic means through which authoritative inscriptions were generated. Promoting the Pitman variety of shorthand in his *Phonographic Teacher* (1852), for instance, Epinetus Webster invoked an admired technological system of his day:

What the steam engine has done in locomotion and commerce Phonography will do in fastening thought upon paper.[21]

He particularly commended "this mental railroad" to reporters, lawyers, ministers, and editors, assuming, it seems, that these men had greater, more important needs than others to "fasten thought upon paper." His railroad metaphor, like the "information superhighway" of today, neatly conveys a sense of necessary infrastructure, possessing breadth, connectivity, and access, while it helps defy awareness of exclusion, obscurity, or the loci of power and capital. On the basis of this kind of thinking, court reporters were the first target market for the original Remington typewriter when it was marketed in 1874, and a number of court reporters later dropped out of reporting to sell phonographs and graphophones in 1889, impressed by their potential. Rule-based shorthand and the two inscriptive devices were together enrolled within emerging forms of mechanical objectivity.[22] After court reporters, both machines were marketed and sold to others who stood in particular need of valid inscriptions—businessmen, clergymen, journalists, authors, and anthropologists, men who comprised the culturally authoritative witnesses of the age. Psychical researchers meant to be a part of such august company. Some among them even drew confidence from the benign interests of differently expert witnesses, like Edison and Sir William Crookes.[23]

Typewriters of various sorts had been invented since 1713, but the machine developed by Christopher Latham Sholes and made by the Remington Company was the first commercial success. Three months after buying his typewriter, Mark Twain was persuaded to write a testimonial for Remington, which the company milked for years in its advertisements:

Gentlemen,

Please do not use my name in any way. Please do not even divulge the fact that I own a machine. I have entirely stopped using the Type-writer, for the reason that I never could write a letter with it to anybody without receiving a request by return mail that I would not only describe the machine but state what progress I had made in the use of it, etc., etc. I don't like to write letters, and so don't want people to know that I own this curiosity breeding little joker.[24]

Like the phonograph and all inscriptive inventions, the first thing the typewriter did was provide evidence of itself (see Chapter 4 of this volume); Twain's typescript letters bore witness to Twain's possession of the machine. In private Twain compared typing to typesetting, with which he was greatly familiar. It reminded him, he wrote to his brother, of a typesetter they had known, who "used to set up articles at the case without previously putting them in the form of manuscript." Twain's admiration for this typesetter, who composed as he composed (who set type as he thought up his arti-

The Remington.

Light Touch

Durable

Shift Key

Good
Manifolder

Simple

Manufacturers' Price, $122.50.
Our Price, from $35.00 to $75.00.

FIGURE 16. "Invisible" writing. An upstrike Remington typewriter.

cle), resembled his admiration for the machine, which seemed to require the same thing.[25]

Twain's analogy made particular sense given that the early Remington was an "upstrike" machine and did not allow its operator to
see the composed line until after it was typed. Like the printer's
compositor, the typist could not see his or her printed work as it
progressed. Since the letters were printed against the bottom of the
platen cylinder, the typist could only see them after several more
lines were typed out, unless he or she stopped typing and "raised
the carriage up on its hinges and peeked" (Masi, 29). The Remington typewriter was a black box: work entered, product emerged.
What happened in between may have seemed mysterious to some
and self-evident to others, but it was not a public or a human matter, only a secretarial and technological one. The machine's upstrike

design seemed to refute the possibility of error, however unrealistically, and in removing the act of inscription from the human eye seemed to underscore its character as a newly technological and automatic event. Here was a new form of "sitter" and a new "invisible." The arrangement would prove metaphorically potent, even after later makers of typewriters abandoned this design.

Scholars have overdrawn the history of the typewriter as a triumphant progress from the upstrike Remington to the "front strike" Underwood of 1894, which had "visible" writing that allowed the typist to see the letters as they were being typed.[26] *Visibility* was long a central concept in the context of typing, though only after the introduction of the Underwood did the term itself consume the industry, taking its place with the ubiquitous adjective "standard" in typewriter brand names and promotional literature. In advertising, some mention of visibility seems long to have been obligatory. The Hammond Company, promoting a typewheel machine it had introduced in 1884, claimed all "Work in Sight." The Daugherty went to market in 1890 with the imperial slogan "We Claim Everything 'In Sight'" and as late as 1912, the Carona Company felt it had to mention that the Standard Folding Typewriter had "visible writing." Even Edison's Bates Manufacturing Company dallied with a "visible indicator" numbering machine during 1910. The Columbia Typewriter Company boasted in its *Psychological Review* advertisement that the Bar-Lock machine typed "every letter in sight of operator." The claim must have seemed necessary, even if it was not strictly true, given the "top strike" design of the machine. The Bar-Lock operator could likely catch a glimpse or two of the page through the typebars and decorative grillwork, but he or she would really have to be standing over the machine to see the letters as they were typed. Misleadingly pictured in the advertisement, "VIEW AS SEEN WHEN WRITING," the height of the Bar-Lock typebars and grillwork is greatly foreshortened. With such mendacious exaggeration unavailable (because of its product) and possibly unappealing, the Remington Company countered in its advertisements for the upstrike machine that good typists didn't *need* to see the letters as they were

FIGURE 17. Bar-Lock typewriter. *Left*, Bar-Lock typewriter advertised in the *Psychological Review*. The Bar-Lock had a double keyboard, QWERTY and qwerty. *Above*, side view. The Bar-Lock was claimed to have "visible writing," but its operator probably had to crane her neck to see the typewritten text. (Jersey City Public Library, Jersey City)

typed. This was a truth based on the spread of the "touch-typing" technique and harkened back to debates over shorthand and just how involved the anatomy of the stenographer was supposed to be, exactly what part head and what part fingers. Also called "all-finger" typing, touch typing vied with "visible" typing in the construction of typing as a bodily experience, aligning it more with the sense of touch or more with the sense of sight. Inscription, in either case, might have little to do with the mind.

The odd terminology of "visible" and, by implication, "invisible" machines, and therefore visible and invisible writing, attests in part to a lingering confusion between the machine and its function and in part to the displacement of contemporary assumptions about handwriting as a graphological trace, revealing of character, gender, class, and education. Part of the process of typing, not the machine, its output, or its author, became newly visible with front-strike machines like the Underwood. The word *typewriter* itself was the site of a similar confusion. For more than twenty years it referred to both the machine and its operator. The author of an 1895 *Atlantic Monthly* article entitled "Being a Typewriter," clarified glibly that she had in mind "the human being, and not the machine."[27] The word "typewriter" was also used as a verb, meaning "to type," and typescripts were said to be "typewritered." These lingering, linguistic confusions between the device and its function, between the typist, typing, and the typewriting machine, indicate the lengthy negotiation of typing as a modern activity and typists as a labor cohort. Such confusions were eventually dispelled, even as the typewriter underwent its lengthy construction as a commercial product and a design object. (Between 1874 and the 1920s, there were three hundred different brands of typewriter, but after a volatile early history and a rash of corporate mergers, there were only five models being made in the United States by 1929.)[28]

Grammatically speaking, a familiar noun, "type," had finally been adapted into a transitive verb, "to type" (like "to fax" in the 1980s), while the intransitive and transitive qualities of inscription remained less fully determined in the rhetoric of "visibility." Gertrude Stein had hoped that "character" would be revealed in automatic writing, while contemporary graphologists held that "true character" was revealed in any handwriting. Inscription in these instances held no special or essential character as important as its permanence; "Writing is nothing but a number of movements," wrote one graphological expert, "which have become visible immediately after being made, and which remain so" (Hagen, 2). Even those less sure of graphology, which like phrenology or psychical research long vied for scientific status of its own, knew the ways that handwriting styles might be revealing of the gender, class, and education of their writers. In a certain sense the rhetoric of "visible" typing partly replaced a transitive ("Writing reveals character") with an intransitive ("Typewriting is revealed") logic of inscription.

The same intransitive logic tellingly challenges the norms of industrial design. Most modern mechanical devices have progressed in the opposite direction from the one taken by typewriter design, moving toward a black box and away from visibility. Phonographs, once open, tabletop hunks of machinery, became popular in cabinet models, and after the introduction of the Victor Victrola in 1906, even the phonograph horn was built into the furniture. The design of machinery as diverse as automobiles and toasters has followed the same trajectory; housing removed the working parts of the machine from the vision of its operator. In its retreat from the eye, modern machinery evinces its own acceptance and familiarity, the accomplishment of its transformation from invention to commodity. Inventions are component whereas commodities are surface. Indeed, even typewriter manuals that once introduced the "operator" to the internal mechanics of the machine soon offered her only maps of the keyboard.[29] That the preeminent functional aspect of the typewriter went the other way, from black box to open box, suggests cultural needs as well as assumptions about writing, about how typing was automatic, distinct from some technological functions and comparable to others. The appeal of "visibility," the eventual and, in part, commercially constructed necessity of seeing into the process of typing suggests that the typewriter is a tool, like Remington's other products, the sewing machine and the gun, or like a pencil, functioning at the expense of directed attention, labored *with* rather than upon. Unlike the phonograph, which emerged from the laboratory as both subject and object, both the means and the end of technical inquiry, the typewriter seems to have settled into the marketplace as neither subject nor object in those terms, rather as a minimally visible intermediary between manuscript and typescript, author and page. Like the pencil according to Henry Petroski, it be-

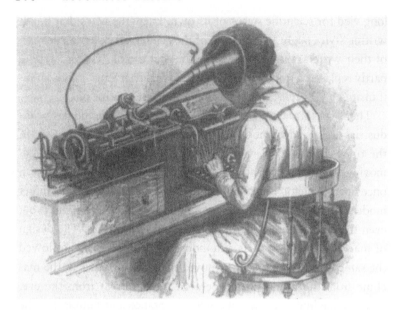

FIGURE 18. Modern media. Woman worker mediates between dictation phonograph and typewriter (1892). She is working on an upstrike, "invisible" machine.

came part of "what we forget." Though it directs attention, the business typewriter itself became crucially unseen and unremarked upon. Skilled typists do not look at the keys or at the body of the machine; rather, they experience what William James undeniably would have called a "cultivated motor automatism" or a "mild case of possession," when they cease to exert intention before every action of hitting the keys. And admittedly, if the "visible" typewriter became gradually less visible, unnoticed, then the same fate befell the typist during the early twentieth century. She receded from view, present occasionally in a few lowercase initials at the bottom of a business letter and noted, in the representations of others, as a mere tap-tapping sound that permeated the modern office, evidence of work in progress as surely as the tap-taps heard during parlor seances spelled out connections between the living and the dead.

～

I am suggesting a connection between spiritualism and typing based upon more than historical coincidence. There was a twofold similarity between these activities of cultural production, a similarity that depended upon related literacy practices. Both were sites of anxiety about visuality and textual evidence and injected gender relations into complicated authorial acts. In spiritualism and psychical research, communication with "invisibles" was registered differently through the person of the researcher/sitter and the person of the subject/medium. Multiple tiers of inscription provided different accommodations of the same evidence: automatic and nonautomatic writing both vouched for the truth of implausible phenomena. This doubleness was centrally troubled in problematic cases where researchers served as their own subjects, like Stein, but also more generally, due to the disciplinary pressures on spiritualism as a cultural activity bent on believably describing the unbelievable and on psychical research and psychology as new academic fields in which and for which individuals sought recognition. In their zeal to prove themselves, psychical researchers embraced the automatic nature of shorthand as a means of establishing the automatic writing of hypnotized subjects. Typing too emerged with force as a multitiered relation, so often separating the businessman-author of a text from its mechanical inscription. Like the literary author or journalist, increasingly alienated from the mechanics of the printing press, business correspondents and other writers became increasingly divorced from the mechanics of producing their own authored texts.[30] And like the sitter and his medium, the businessman and his typist based their relationship partly on his matter—writing about him—and partly on "invisibles," in this case on the letters, words, and lines of a typescript that appeared unseen or unwatched on the page.

Both spiritualism and business typing were consumed and, ultimately, seduced by matters of visuality. Seance participants spent their time trying to see "invisibles" in a darkened room—an unnoticed paradox—while the manufacturers and purchasers of typewriters seem to have been sold on "visibility" because of the nature of writing, not typing, or at least because of the nature of error, not success. Whatever their complications, both pursuits demonstrate a preoccupation with method, the more technical the better, as a way to vouch for content. Spiritualism ached for such a method, and typing seemed to be one.

The problem of both spiritualism and psychical research was the problem of locating authentic authorial agency. Was the medium really a medium for authorial spirits of the dead? Or was she, or was her male handler and promoter really the author? Nathaniel Hawthorne, Henry James, and William Dean Howells were fascinated with the problem in their respective works *Blithedale Romance*, *The Bostonians*, and *Undiscovered Country*, and so were American readers. One of the best sellers of 1894–1896 was George DuMaurier's novel *Trilby*, serialized in *Harper's Monthly Magazine* between January and August 1894, and then published as a book. In less than six months, *Harpers* had sold 200,000 copies of the anti-Semitic novel, a sentimental tale of hypnosis, and America experienced a "boom" of Trilby reading, Trilby drama, Trilby spin-offs, and Trilby parodies.[31] Set largely in Parisian bohemia, the novel centers upon a laundress with loose morals but a good heart, who falls under the hypnotic thrall of Svengali, an evil, Jewish, musical genius. Though tone-deaf, Trilby is made to sing in a trance, and the result is perfection, a phonograph, "a singing machine":

> Her voice was so immense in its softness, richness, freshness, that it seemed to be pouring itself out from all round; its intonation absolutely mathematically pure; one felt it to be not only faultless, but infallible; and the seduction, the novelty of it, the strangely sympathetic quality! (DuMaurier, 248)

> [Svengali] had but to say "*Dors!*" and she suddenly became an unconscious Trilby of marble, who could produce wonderful sounds—just the sounds he wanted, and nothing else—and think his thoughts and wish his wishes—and love him at his bidding with a strange unreal factitious love . . . just his own love for himself turned inside out—*à l'envers*—and reflected back on him, as from a mirror. (357; ellipsis and emphasis in original)

Trilby's artistry is an impossible blend of perfection and artifice, made more romantic in its inarticulate and inarticulable musicality. She is neither the author of her self or of her voice; Svengali is. And it is partly her lack of intention, the invisibility of authorial agency, that captivates her audience. Listeners witness a spectacle of origination within which origination itself is elusive. Other professional or aspiring artists throughout the novel simply possess or lack natural talents, while Trilby's talent is unnatural or is no talent at all. Explicitly "automatic" writing does not appear in the novel, though the narratorial voice intrudes every once in a while with diffident comments on the conventions of sentimental fiction, and DuMaurier himself played with the same issues of authorial talent and agency in interviews about the novel. He claimed that the story had come naturally to him; "Of the mechanism of the story, I can tell you nothing. The story formed itself" (qtd. in Purcell, 62).[32] Flippancy perhaps, but it was precisely this possibility that troubled emerging identifications of sentimental pulp, lowbrow, or formula fiction—from which Trilby eventually suffered—as well as spiritualism and typing.

The typewriter shared the dilemma of indeterminate authorial agency. If handwriting harbored personality or character as well as the telltale signs of gender, class, and education, then the typewriter harbored only an opaque "standard" writing. Graphologists would be frustrated, but so would every reader. For a long time typewritten correspondence was considered impolite, no matter what it said. (Etiquette still requires handwritten notes or letters in specific "social" circumstances.) Business, of course, has never really been polite, and in it typing prospered. Still, an anxiety remained. There were signs that typing might obscure authorship, particularly its personal, human quality, by obscuring the true character of the author. Praise for the machine hinted as much, as authors suggested that typing might improve the style of written English.[33] As early as 1889, a woman contributor to The Writer magazine urged:

There is much to be said in favor of the type-writer as an aid to fluent composition. There is with most writers a nervous strain, consequent upon the inability of the hand to keep pace with the rapidly-flowing thoughts, which is far more wearing than the mere fatigue of the hand and the arm. (McCray, 114)

Here was the "nervous, high-strung, very imaginative," Type One person that Stein would later describe. Her very character seems liberated from the necessity of expressing itself in newly mechanical and "fluid composition." This might be a boon to her, but her words would also be less authentic expressions, less true to character. Her typewriter was an "aid to composition," but it composed its writer into someone new, a self-forgery.[34] And in another sense business typing involved an opposite and simultaneous risk, that the typescript might harbor the personality of its typist rather than, or in addition to, the intentions of its businessman author. "Personality" or "character" might be a desirable trace of authenticity in literary authorship, but it had no place in business, as dictaphone salesmen seeking to replace office stenographers made clear. They assured the "dictator," as they always called the businessman author, that with one of their machines, "There is no personality to distract, no human limitation in recording correctly the concentrated efforts of a dictator."[35]

If the writers of shorthand manuals and the publishers of shorthand reports had long found themselves bereft of copyright, competing in a marketplace without statutory authorship, and if the makers of phonograph records found themselves uneasily relying upon patent rights to protect their now-readable commodities, the users of typewriters converted such dilemmas into the vernacular. Literary authors, correspondents, and businessmen all lost the habitual, graphical appearance of authorship and with it some of their accustomed rhetorical authority as authoring agents. Evidence that typing involved new anxieties about the location and character of authorial agency, the authenticity of authored texts, as well as the gendered conditions of their production emerged in popular literature about typing as well as in the rhetoric promoting the "noiseless" machines that soon joined "visible" ones in the marketplace. In popular literature the author's character always managed to emerge from the authored typescript. Starting in 1891 with an early adventure of Sherlock Holmes, the character which usually emerged from the authored typescript was a defective one: "It is a curious thing," remarked Holmes, "that a typewriter has really quite as much individuality as a man's handwriting. Unless they are quite new, no two of them write exactly alike" (Doyle, 256).[36] Holmes deduces the identity of the "disappearing bridegroom" by noticing the "trivial but characteristic defects" of his typescripts. Holmes's method, based on the forensics of typewriting, became stock and trade in detective fiction, and appeared in other genres as well. In O. Henry's "Springtime a la Carte" (1904), a lonely typist from the country is discovered in New York by her beau when she inadvertently (automatically?) types part of his name into a menu she prepares. If this were not enough, he is able to recognize a characteristic flaw in her typescript font. As in the earlier story there is a conjugal interest in the typescript, a normative romantic relationship that is either subverted (by criminal psychopathology) or achieved (by unbelievable coincidence). In both stories the plot revolves around "flawed" or unconsummated gender relations and "flawed" typescripts. While gender symbolism like this did not appear in every fictional account of typewriter forensics, it does provide a reminder that typewriters and the business offices that contained them were sites of radical change in the gendered nature of work and in the public relations of men and women. Business and public life were "engendered" as never before, as Angel Kwoleck-Folland and other scholars have explained.[37]

The grandest of all the early fictions of typewriting is Bram Stoker's *Dracula* (1897). Though Mark Twain claimed the distinction of having submitted the first typed manuscript to his publishers, Stoker's novel is in all likelihood the first to fictionalize itself as a typescript. On the final page of the novel, readers learn that the text they have just consumed contains "hardly one authentic document; nothing but a mass of typewriting" and a few bits of handwritten matter. "We could hardly ask any one," Jonathan Harker muses, "even did we wish to, to accept these as proofs of so wild a story" (Stoker, 382). The mass of typewriting has all been compiled by Miss Mina Murray, pieced together from shorthand reports, dictaphone transcripts, printed telegrams, and other documents. But in the printed novel, the diverse materiality of these fragments is largely hidden from the reader, who consumes vicariously, vampirically, without experiencing the fictional lack or accumulation of authenticity and proof. Mina Murray, whose name so echoes that of Thomas Edison's second bride, Mina Miller, is one of Dracula's victims. Jennifer Wicke reads the novel as a self-contradicting tale of "vampiric typewriting," a tangle of production, consumption, and distribution that critically glimpses the emerging mass culture in its representations of tabloid news and international capital. Like its contemporary, *Trilby, Dracula* hinges upon invisible yet sexualized authorial agency in a world that is otherwise marked by certainties of gender, class, race, religion, and the provincial. The mystification of agency against the comforts of such certain and customary parameters of identity forms the appealing or thrilling "noise" of these novels, a metaphorical noise directly related to the real noises of typing.

Typewriters, like phonographs, involved sound in the process of inscription. And like shorthand, both machines were buffers between aural experience and inscribed fact. The obscurantism of shorthand alphabets and the hieroglyphics of phonograph recordings were solved by the "standard" appearance of typescripts, which nonetheless involved such a recomplication of visuality in their "visible" textual production. The dictator's voice, made scientifically phonetic by the Pitmanic reporter, acoustically and indelibly reproducible by the phonograph stylus, was itself echoed in the inarticulate clatter of typewriting. Unskilled typing had its own profane noises, which Twain observed as "the kind of language that soothes vexation," but skilled typing had only the mechanical noise of letters, marks, and spaces hitting the paper in sequence, punctuated by the arrival of every new line of type as the carriage was returned. The pen scratching across the page was negligible by comparison; the clatter of the printing press outside the experience of most individuals. Spiritual telegraphy and music remained the most viable points of comparison. Various early models of the typewriter and its elder cousin, the printing telegraph, had been formed with piano keyboards instead of multiple rows of smaller keys. After all, the piano was the most familiar keyboard instrument before the typewriter. Much later, commentators such as George Carl Mares still called the typewriter a "literary piano," when groping for an apt description. Touch, or all-finger, typing confirmed the comparison, since the eyes rest "not on the keys but on the copy, as the eyes of the pianist rest on the score."[38] Unlike the piano, however, the typewriter was far from sonorous, though Henry James wrote of being conditioned to the "music" of his secretary's Remington.[39] The "noiseless" machine was thus a premium at the same time the "visible" one was, and like the concept of the latter, the idea of noiselessness possessed a certain plasticity. No machine was silent, no matter how well engineered. Yet many thousands of "noiseless" typewriters were sold. The liberality with which the term was used and accepted suggests more than just commercial zeal and consumer apathy. It evokes the "technopoetics" that Strother Purdy describes, "the *fantasy* of the utterly decreet, silent, dependable servant," seen and not heard, which hovers behind so much modern machinery (136; emphasis in original). Automatic *women* are the relevant objects of desire.

Like the ubiquitous term *visible*, the term *noiseless* attests to the degree that matters of control, of eyewitnessing and evidence, inhere in the gendered, symbolic meanings of automatic writing in the modern business office. The sound of business was the clatter of nearby typing. But the typing was "noiseless" in the sense that it was usually "only" a medium, a handy device for the material realization of the dictator's voice.

The paradox of noiselessness mirrored the paradox vital to psychical research of believably describing the unbelievable. Automatic writing in both contexts was mystified by the interval between authoring and inscribing, an attention gap that could variously contain authoring agents and inscriptive means. In spiritualism, invisible, tapping spirits of the dead inhabited the gap. In Gertrude Stein's psychological research, it was the automatic writer's character. A little later it was the Freudian unconscious that surfaced in the interval, which gave automatic writing its appeal to the surrealists. Indeed, the Muse was one traditional visitant in the same space; Goethe and

Yeats each produced extensive automatic writings as a part of their oeuvres. And the typewriter reportedly became a sort of object-muse, a fetish, in the creative processes of Henry James, William Carlos Williams, Ezra Pound, and many other literary authors. Stein continued profitably to explore automatic writing of the secretarial sort. She rather notoriously made Alice B. Toklas her typist, part muse and part drone, conspirator and audience, the combined subject and object of autobiography. In business typing the interval between authoring and inscribing became cluttered with a variety of possible means, the stenographer, dictaphone, or tentative rough draft that joined the skill and gender of the typist, the engineering and design of the typewriter. Typing didn't create the gap, it only called new attention to it, prolonging and stylizing it in new ways, complicating intention, adapting and supplementing the logic of the hand and pen and telegraph, facilitating more and different intrusions by the white-collar labor market, gender, and related mysteries about authoring. Visibility and noiselessness became twin components of modern textuality, ever harboring their own undoing in their respective and defining opposites—invisibility and noise. The newest immediate aspect of such textuality was probably the sound of blank space, as the spaces between words and lines of type had to be created, rather than simply "left" blank as they were in the production of handwritten pages. In typewriting, space on the page was *made* as well as *used*: writing newly involved "writing space."[40]

Coda:
The (Hyper)Textuality
of Everyday Life

It happened that a gentleman had an idea: if all handwriting preserves within its fly tracings the character and therefore the destiny of a person, like the grooves on a disc when you know how to be a stylus, then you would only have to write exactly like Napoleon to set in action the inverse process and produce, not a Dvořák quartet brought to you by a spiral and a little motor, but Bonaparte, who at this very moment, having completed five months of handwriting study, leaves his house on the rue de la Convention and doesn't go a block before he is seen by four streetsweepers going about their tasks, and the legendary magnetism of the Eagle of Austerlitz overwhelms not only the streetsweepers but also a woman selling eggs in a doorway, instantly transforming them into soldiers, not to mention a number of priests, three masons, and the salespeople at the nearest hardware shop, all of whom rush to line up behind the Emperor, so that a small but select and, most of all, fervent troop advances along the rue Vaugirard as several dumbfounded neighbors contemplate this spectacle from their windows and discuss it, shocked and horrified, until the moment the Emperor raises his eyes and lifts his arms, and then the first bits of paper about the great triumph rain down all around him, since we are, after all, well into the second half of the twentieth century and the outward signs of idolatry and celebration have changed considerably, thanks to the widespread democratic customs of the Americans.

—Julio Cortázar, "On Graphology as an Applied Science"

Why do so many recent accounts of hypertext and the reading and writing associated with computers neglect to mention the developments of 1877 to 1914, when so much other scholarship identifies that period as a crucial point of rupture and transformation? I want to underscore the present relevance of the inscriptive devices I have been describing, but in order to pose this initial question responsibly let me first indicate the ways that Richard Lanham, George Landow, and David Jay Bolter, three critics who have published books on digital textuality and are seen by many as the spokesmen of a new age, each use the past in order to describe the present and posit the future. While there is much to admire in their accounts, I am troubled by their only slightly varying positions. Each argues that digital hypertexts "democratize" reading and will, in turn, democratize education and may well democratize the social order. This rank determinism, reminiscent of Ian Watt and Jack Goody's earlier position that literacy forms a cognitive precondition of democracy, seems to me to reside less in what these critics have to say about experiences of textuality than in the way they use historical explanation. I am, in part, quibbling with the rhetorical device of overstatement: Landow writes bluntly, for instance, that the Western availability of relatively inexpensive writing surfaces "led, around the year 1000 [C.E.], to interword spacing," which in turn "permitted reading silently, which, in turn, led to our modern notions of a private interior self" ("Twenty Minutes," 217). Thus reduced to one sentence, history is shockingly reductive. Part of the effect is to make present-day structural and phenomenologic innovations resonate with the "revolutionary" power that a few other developments in the history of textuality appear to have had when they are viewed with a lot of accumulated hindsight. For these scholars, the revolutionary point of comparison is usually Gutenberg's moveable type.

The history of computing should be distinguished from the prehistory of hypertext. Scholars who narrate the history of computers usually start with the calculating devices of Wilhelm Schickard, Blaise Pascal, Gottfried Wilhelm von Leibniz, and Charles Babbage. Add to this the partly bureaucratic need for repetitive calculations, in the U.S. Census, for instance, or in business, artillery training, and cryptography, and you get ENIGMA, ENIAC, IBM, and UNIVAC, with Apple, Intel, and Microsoft not far behind.[1] The prehistory of hypertext is, by contrast, the history of textual and graphic forms (particularly the printed page), which "contain" the information people need as social beings. Lanham, Landow, and Bolter each point to the inadequacy of print as a container for information by noting either the textual experimentation of canonical authors, like James Joyce, Lawrence Sterne, and Jacques Derrida, or by appealing to historical analogies. They draw comparisons to biblical exegesis, with its progressive encrustation of marginal additions, intra- and intertextual allusion, or to a futuristic "Memex" encyclopedia system imagined by Vannevar Bush in 1945, with its interlinked "trails" of inquiry connecting units of information.[2] In all of these cases, pages of print prove their limitations as what Bolter calls "perceptual" forms. Authors and readers are rankled by the limiting linearity of print, the single sequence of finite pages within a book or the single sequence of books on a finite library shelf. These structures constrain the organization and consumption of information.

Geoffrey Nunberg and Paul Duguid have both examined the assumptions behind this historical account of information as a "uniform and morselized substance" that can be liberated from the bondage of print by the links and multimedia of hypertext (Nunberg, 116). Duguid proposes a counterhistory of hypertext that begins less with the Gutenberg revolution of the fifteenth and sixteenth centuries than with the nearly simultaneous invention and appropriation of double-entry bookkeeping. Bookkeeping works by forging and perpetuating links between texts, media, and representations.[3] For five centuries readers have jumped between mutually related parcels of information, each meaningful in and of itself as well as meaningful within a material and semiological system of economics, accountancy, and trade, the circulation of value, paper, and goods. The accountants' general ledgers (organized by account) and the journals (organized chronologically) are together a kind of paper mainframe, helping to process information in parallel and in series. These functions exfoliate bibliographically and geographically into a network of specialized texts, like the detailed ledgers of a particular account, or the daybook kept behind the counter to journalize sales, or kept onboard ship to journalize lading and disbursements. Into the columns of such books impinge different inscriptive forms, like shipping receipts, checks, and bills, which acquire part of their meaning in the consenting practices of their exchange and part of their value in the consensual practices of their circulability. This second prehistory of hypertext appeals to me because it acknowledges the numerical and bureaucratic genealogy of computing. At the same time, it endorses a vernacular, or everyday character, of inscription and books that is less encumbered by the "Cultural" status of texts now so difficult to disentangle from presence and use. A corresponding prehistory of computing might helpfully complicate the numerical biases of standard accounts, by noting and including the elaborate search and retrieval architecture, for instance, in places like the main New York Public Library (1907), with its own alphabetic (not alphanumeric) cataloguing system for warehoused books and its infrastructure of pneumatic tubes and dumbwaiters for sending call slips and receiving texts. Or it might consider the integrated structure and semiotics of Grand Central Station nearby (1913), with its routes and signals for trains, its routes and signals for passengers, and the tiny spiral staircase that connects an information booth on one level (suburban transit) with an information booth on the other (interurban transit).

Substituting a different prehistory for hypertext does much to challenge the novelty ascribed to digital textuality by Lanham, Landow, and Bolter. My point is not to deny the exciting potential of digital communications, or to claim cynically that there's nothing new under the sun. Rather, I want to question and elaborate the parameters of novelty that recent accounts of hypertext seem to posit as the foundations of a new democratic future. Keenly felt should be the gleeful claims originally made on behalf of shorthand, phonographs, films, radios, and televisions, each supposed to harbor democratization in its own way.

At least for now, Nunberg admits, the "[World Wide] Web is still the electronic equivalent of a gated suburban community" (133).

This means that it is full of a lot of confusing, banal (or worse) "suburban chatter," but also that its access is restricted to those who either possess the necessary economic resources or can politely associate themselves with the right institutions of education or employment. It also means that once inside the gate, there are fences in the distance that might be hard to see except in snatches, but that hem one in nonetheless. Karen Ruhleder finds that classicists using the new on-line concordance of ancient Greek literature have a different critical sense than the classicists using multiple, printed editions and bound concordances. Their perspectives differ because, "In direct contrast to a textual edition with explanatory notes," the "complex decisions about the construction" of the hypertextual databank remain "hidden from view because of the manner of its presentation" (53). In this case, the *Thesaurus Linguae Graecae* seems to be reorienting the way an entire academic discipline thinks of itself, the kinds of questions that get asked, and the methods of study employed and valued in answering them. Yes, younger and far-flung scholars are now able to troll through all of ancient Greek literature, to compete with the bibliographic experience and resources of senior scholars. But the editorial decisions to discard some textual variants and digitally enshrine others have all been made beforehand by a disciplinary panel of experts in an editorial process nowhere present on the screen. Calling these new and different fences "democratization" seems simplistic at best. A narrow example, perhaps, but my contention is an obvious one. Reminder: systems are designed to obscure their own limits. "Interactivity" is only active to a point, since its vaunted flexibility is programmed ahead of time by agents or institutions of power, be they authors and editors or programmers and manufacturers.

For Landow the great promise of hypertext arises in the convergence between technology and contemporary critical theory. Though Landow avoids specifying a cause for this convergence,[4] the implacable progress of technological application and the "instinctive theorizing of hypertext" by Jacques Derrida have together landed "users" at the same place. Structuralism and then poststructuralism, like hypertext, valuably *decenter* "many of our culture's assumptions about reading, writing, authorship, and creativity" (Landow, *Hypertext*, 203). Lanham too writes of a "convergence" among "democracy, technology, and the arts" (Chap. 4). Lanham identifies hypertext with the ancient rhetorical tradition of the Sophists and Cicero (and only by extension the rhetoric of poststructuralism).[5] But he also notes "bi-stable" surfaces and *oscillations* of attention, by which present digital forms have made people more aware of both the way the texts are put together and that put-togetherness has consequences for their meaning. Decentering and oscillation do seem correct and valuable attributes of digital textuality, yet neither is unique, as Lanham's repeated allusion to ancient philosophical quarrels should suggest, or as the varied genealogy of the phonograph should make clear. One of the assertions of the preceding pages has been that all new media, in failure or success, in rejection or in erratic, faddish appropriation, inspire conflicted cultural moments of self-consciousness about the making of meaning. New media *as* new media all inspire a flicker in which the textual and other operational characteristics of old media seem particularly illuminated and, at the same time, decentered and decentering with regard to perception, authorship, reading, and the like. New media variously recycle our awareness of old media. Bolter and Grusin concoct the term "remediation" for what is often experienced as an elaborate "representation of one medium in another" (339, 345), noticed first by McLuhan. Remediation explains why the Web has "pages," as well as why the first radios looked like phonographs and the first televisions looked like radios.

In its latest form, remediation helps explain the vigor of book history as an emerging academic concentration in this age of computers. Hypertext, word processing, optimal character recognition, and other aspects of digitized text have made the traditional codex more visible as a medium and thus more powerful as a subject of study. The linearity of print, for instance, seems particularly stark against the search, format, and linking flexibility of digital media. There is little reason to think, however, that present circumstances grant any more privileged vantage point from which to experience decentering or oscillation itself. Nor should one be satisfied with the Whiggish connotations of the term *remediation*, with its suggestion that new media necessarily improve the old along some path toward perfect, transparent mediation. Might not phonographs and contemporary inscriptive devices have been involved in equally notable disciplinary reorientations, like the relegation of textual studies to literature departments, for instance, and literature itself into "Culture," with that capital letter C? The study of literary texts as such, in the English departments of American schools, colleges, and universities, coincided with the age of emerging mass culture that I have been describing at some length. Perhaps phonographs, cathected with the symbolism of cultural difference and potent with the decentering of texts, authors, and readers, helped enable precisely the disciplinary cartography of the humanities that seems so in jeopardy today to some observers.[6]

Experiences of textuality are as variable as they are fluid. The new inscriptive devices of 1877–1914 helped complicate writing and reading in ways that depended upon social practices as diverse and variable as the perception of racial and ethnic differences, the negotiation of gendered workplaces, the creation and perpetuation of public memory, the apprehension of modern machinery, and the organization of mass markets. One benefit of historical comparisons should be to acknowledge this breadth rather than to obscure it.

Like hypermedia today, early phonograph records were the subject of great approbation and much speculation, some of it contradictory. Records provided small amounts of machine-readable text, out of an infinite and heterogeneous body of information. With a little more research and development the amount of memory they possessed could be increased dramatically. The cost of memory would therefore decline, and more data could be stored. Records were hailed for the immediacy and quality of their retrieval speed, which was said to approach the ease and flexibility of oral communication. They featured a read-only function as well as the ability to receive new information, sharing "with the wax tablet the quality of rapid change" and thus "promoting an ideal of writing very different from the ideal of print," as Bolter writes about computing (*Writing Space*, 40). Using these new texts could be a solitary activity, or it could involve many users simultaneously. A new kind of "reading" would be more natural, closer to the way the human mind really works. Mailing someone a record of your voice, while hardly an instantaneous communication like telephony, would provide a new form of networked information, sound itself, easily downloaded from one phonograph, saved, and uploaded on another. Education could be made more fun and interactive. Schools using phonographs to teach foreign languages boasted of interactive learning, while their students had to learn quite a lot about hardware and interface conventions before they could learn anything about Spanish or French. Recordings could discipline and instruct, but they could also inform, in wax "newspapers" and public announcements. They could entertain (music) as well as pacify ("Muzak"). Public taste would be elevated as working-class and rural audiences could be exposed to opera and the classics. Democracy would flourish as constituents could really have access to the speech acts of government. Two different operating systems would compete for market share, cylinder versus disc. Consumers would then have to worry about incompatible file formats, since lateral-cut gramophone records and vertical-cut phonograph records could not be played on the same machines. Success would go not to some "best" machine hailed by engineers, but rather to the manufacturer who was able to drive sales with software as well as hardware. It would emerge out of the building of market commitment while perceiving and adapting to changing conditions in the structure of sales and distribution, intellectual property law, and the character and intensity of competition by clones.

Like the World Wide Web, silent motion pictures nested graphic forms within small blocks of text—titles, labels, dialogue, and credits helped them make sense. The result was a "breakout of the visual," as Bolter writes of digitally influenced forms, in which "we can discern a struggle between textual and visual modes of representation" ("Ekphrasis," 258–62), within which the primacy of graphic signs seems assured but for which a textless future is more of a defining impossibility than a goal. Audiences remain readers, even when they "see," rather than read, a movie, even when they walk out before the credits scroll by. Silent films were available at an international network of specific sites and quickly allowed multi-user access. At first the companies that made hardware also made software, but corporate reorganization gradually separated manu-facturing and production, with most software writers and produc-ers clustered in one area of California. Self-censorship and then cen-sorship sometimes modified content. Ratings modified access. News and educational functions vied with amusement, and audiences had to become attuned to whole new conventions of facticity and artis-try, standards of evidence, and parameters of quality. Intellectual property rights were used by producers to stabilize distribution net-works until antitrust litigation came to the aid of so-called "inde-pendents." The standards of data storage were quickly arrived at, though data transfer was eventually needed as the silver nitrate stor-age format was improved upon, and then as monochrome displays gave way to color. Motion pictures that once filled a whole, huge room, later became available for personal, home use, where they fit neatly on a tabletop.

Like the Internet in the 1990s, radio began its nonmilitary exis-tence as a noncommercial anarchy, a freely evolving space of en-coded, verbal expression and shared bricolage or tinkering for a growing community of amateurs, each huddled over his own set. ("His" because their activity was a gendered one.) Like telegraphy or telephony, radio connected separate users, but its unfamiliar "wireless" quality made its connectivity virtual in effect. Real space, articulated in a dense carpet of telephone wires, telegraph circuits, and transportation networks, was newly offset by virtual space, ar-ticulated only by blips or voices in ether, antennae, and Hertzian waves. Radios were emphatically noninscriptive; they "dematerial-ized" writing, just as Bolter says of computing. Formal and informal voluntary associations of "hams" resisted the commercialization of the medium. As Daniel Czitrom and Susan Douglas describe, many users during the early 1920s were appalled at the thought that radio would become a passive entertainment, something you listened to, instead of something you did. And not unlike so much Web browsing today, the sport of "DX-ing" involved trying to make the furthest connection possible, seeing what was out there, rather than becoming engrossed in the substance of any one site on the dial.[7] While a couple of big companies controlled the industry, private users cultivated talents for intercepting signals between unknown parties and at receiving signals from distant stations or from the authorities. They made up new identities for themselves, pretending to be who they were not. They hacked into forbidden parts of the broadcast spectrum, upgraded their sets when they could, and spent their leisure hours chatting, checking the news, lobbying against regulation, and comparing notes on hardware and system operations. International disaster at sea helped galvanize public attention toward the reliability of the new medium, with the *S.S. Titanic* fulfilling the tragic role of TWA Flight 800.[8] Despite controversy, many vaunted the medium's value for public service, as thousands of amateurs could pass along distress signals or civil-defense instructions in times of trouble. And as commercial stations emerged (after wavelength was recognized as a new form of real estate and call letters as a new form of address) many supposed enlightenment and democracy might increase, due to required educational programming, highbrow entertainment, and the potent immediacy of information, news, and public life.

One point of these comparisons is that comparisons are easy. Surely the typewriter was a desktop-publishing system. Perforated piano rolls were the software of the player piano. And each new shorthand alphabet vied to be a kind of human ASCII. Even the gliding planchette used by spiritualist automatic writers (and still familiar to Ouija board users) can be glimpsed in the mouse or pointing device that controls the cursor on a computer screen. No one level of comparison should stand apart, however. Media are textual, graphic, structural, visceral, commercial, public or private, owned or shared, centering and decentering in a host of interconnected ways.

The symmetries that a planchette and "talking board" share with a mouse and motherboard cannot be adequately understood without teasing open the automatic writing of the 1890s, when the Ouija board, the Remington, and the Underwood were all for sale.[9] If the connection makes sense because of the complicated physical and mental experiences of hands gliding and pointing, then what must these experiences mean with regard to the dynamic contexts and constructions of "normal" authorship, writing, work, and sense? How are these experiences themselves shaped? If one remains satisfied with quick, flat comparisons, then one is letting technology explain more than it should, and more than it can.

Experiences of computers as textual devices, like the experiences of phonographs or silent films before them, must depend upon much more than the eventual habits of what are now called *end users*. Experiences of digital textuality depend, for instance, upon personal idiosyncracies, like my own kooky ideas of where the data go "into" this beige box and "onto" my hard drive, even as much as it depends upon the realities of software, silicon, and the like. (My unthinking nostalgia for chirographic inscription tries to attach itself to diskettes and dots per inch.) Digital textuality depends likewise on the status of intellectual property law and a whole range of economic structures and commercial organizations, not just the present monopoly of "windows" and the "desktop metaphor," courtesy of Apple and Microsoft, or the present standards of "browsing" "pages" that are somehow "sites." It depends upon shared experiences of where those present standards come from and how changeable they may or may not be. It depends upon the history of keyboards, the spatial relations of letters and commands, upon patterns of error and possibilities for the future. It depends upon assumptions about language and literacy, about the identity of authors and the psychology of authoring, about the subjectivity of reading in its relation to the subjectivities of buying, eating, driving, and all the rest of the things people do.

Reference Matter

Reference Matter

Notes

INTRODUCTION

1. Postmodernists such as Friedrich Kittler and Paul Virilio have been surprisingly romantic in their thinking about technology. I read a lingering determinism in their works, in which they take technological artifacts and innovations as structural features of a symbolic realm—connected to, and evidence of, the vested interests of society and operations of culture but without the encumbrances of complete historical context. This can lead, on the one hand, to a hagiography of inventors as Wordsworthian poet-priests (Kittler calls Edison "the marvelous One" in the introduction to *Gramophone, Film, Typewriter* [14]), and, on the other, to a portentous view of causality within which only certain convenient and isolated historical instances are the "not accidental" expressions of culture—a form of historicity rather than history. None of this is to deny the strengths of such work, and I will return in Chapter 5 to Kittler's project. On the "posthermeneutic," see David E. Wellbery's foreword to Kittler, *Discourse*, vii–xvi; on the "romantic," see Sebastian. Like Goldberg, I doubt that "historic specificity" regarding "the regulation of writing" needs to contradict fully the interests of poststructuralism (7).

2. See Lovett, 3.

3. The SCOT ("social construction of technology") program is outlined in Bijker (1–17) and in D. McKenzie. It is helpfully critiqued by Winner in his "Upon Opening the Black Box." SCOT is in some ways only the most extremely stated of many recent approaches to technology and society, as the subject itself emerges from the shadow of recent social studies of science

as practiced by Steven Shapin, Simon Schaffer, H. M. Collins, Bruno Latour, and others. See Grint and Woolgar. For valuable pre-SCOT glimpses at discourse within technology, see Marvin; within the history of technology, see Staudenmaier. For a succinct overview of the issue of technological determinism, see Smith and Marx.

4. See Boyarin, 6; Street, 9.

5. Paul Duguid gives a cogent explanation of this view and its underlying refutation of accounts of digital communications that distinguish "information" from the media that "contain" it; see his "Material Matters." I have taken very much to heart Duguid's clarion that "looking at communication technologies in the round" avoids "partial, isolated, and antagonistic accounts [about media]. For the book at least, cultural theorists, contemporary bibliographic critics, and literary sociologists have recently begun to do this. It still needs to be done for alternative information technologies" (80). For rich and different accounts of textual materialism, see Gumbrecht and Pfeiffer; Masten, Stallybrass, and Vickers; and Ezell and O'Keeffe.

6. Scholars other than Habermas have differently discerned a public sphere in the same late seventeenth century: Steven Shapin, Simon Schaffer, and historians of science have traced the origins of modern objectivity to Restoration science and its mechanisms (of rhetoric, theatricality, print circulation, and so forth) for the production of public knowledge and an accordingly public space, which is distinct, if triangulated by mercantile capital, the state, and Puritanism; see Shapin and Schaffer; and Shapin's *The Social History of Truth*.

7. See Panchasi. 8. See Reiser, 60–66.

9. See Dickson and Dickson. 10. See B. Anderson, 135.

11. For instance, on the phonograph, see Read and Welch; Frow; and Gelatt; as well as the tireless and awe-inspiring publication efforts of Allen Koenigsberg and APM Press. On Edison, see the authorized Dyer and Martin biography, and the accounts by Milosi; Baldwin; Wachhorst; Millard; and, most recently, Israel.

12. See Weber, 153.

13. See, e.g., Baker, 51.

14. See R. Rosenberg, 21.

CHAPTER 1, *Making History, Spelling Things Out*

1. *Scientific American*, December 1877; see Edison, *The Papers of Thomas Edison*, 3:670–74. The article alludes to the imperfect "Faber apparatus" that has rubber vocal organs.

2. Crary is one good source on the "technique of the observer" during

the nineteenth century. He maps changes in the habits of American vision without the technological determinism of many other accounts that seize upon photography as a preeminent cause. A comparable work on auditory techniques and conditions is Johnson; he focuses on the roles of political and social change, the refinement of cultural high/low hierarchies, and related musical appetites, in its explanation of changing audience behavior between 1750 and 1850.

3. See Martin, 60.

4. See Goody, *Interface*, 55–56. See also Ong, *Orality and Literacy*, for instance. Ong, like Kenner and McLuhan, tends to overdetermine literacy (as he similarly overdetermines technology). See Street for a cogent critique; and also Fabian, "Keep Listening." Olson's *The World on Paper* (7–13) succinctly outlines and qualifies the determinist claims of Ong, McLuhan, Goody, and others.

5. On dictionaries, see Simpson, 142–43; and also Burkett. The nationalist rhetoric of the Worcester/Webster dictionary controversy is rightly tied to other aspects of linguistic nationalism, though publishers of dictionaries and publishers of shorthand manuals are more interested in money than nations. On linguistic nationalisms, see Simpson; Baron; and Looby.

6. United States Circuit Court (Southern Ohio). Graham published his exhibits from the proceedings. Pitman's exhibits comprise the Benn Pitman Papers at the New York Public Library, Manuscripts and Special Collections (hereafter NYPL). Pitman published the bill of complaint, answer, and his own testimony as *Defendant's Testimony*, a copy of which exists in the general collection at the University of California at Los Angeles.

7. See T. Anderson, 212.

8. *Script Phonography Co., Ltd. v. Gregg*, High Court, Chancery Division (1889). All of my research on Gregg was done at the John R. Gregg Papers, NYPL.

9. See McKee.

10. Prospectus, The Light-Line Phonography Co., Ltd., 1889. John R. Gregg Papers, NYPL. Earnings were poor, judging from the fact that Pitman and Graham both continued to take court-reporting jobs. Pitman also tried to invent his own copper-plate engraving process (see *Defendant's Testimony*, 78).

11. Sample agreements, John R. Gregg Papers, NYPL; stereotype plates for sale, $20 each, *The Phonetic Journal* 18 (1859): 86; manual prices from Brown, 284; and from flyleaves for the 1886 edition of Graham's *Hand-Book*.

12. Here I "dematerialize" literacy, to use anthropologist Johannes Fabian's term ("Keep Listening"), the way that custom essentializes the laissez-

faire market. Both gestures are matters of convenience, and I hope they are
not too limiting in the discussion below.

13. Despite the seeming linearity of this chain of events, causes and ef-
fects are hard to parse out; for one full discussion (among many), see Eisenstein's work on the printing press; or Febvre and Martin.

14. The Committee on Standardization of the National Shorthand Reporters' Association found only small degrees of difference in speed and accuracy among top reporters of the Pitman, Gregg, and Graham systems in
1911 (see McDermut).

15. See Hunter, 234.

16. The first-generation computer language is binary code; subsequent
generations are increasingly removed from raw 1s and 0s. Programming
languages such as C are sometimes described as third-generation languages.
Commercial word-processing programs invoke their own generational
progress, Word Perfect 6.1, for example, giving way to 7.0 and 8.1.

17. Given such antiquity, it is interesting to speculate whether some of
the scholars who have so assiduously identified the formulae of orality in
the surviving poems of Homer might unwittingly have identified the formulae of early shorthand reporting. I don't mean casually to disparage the suggestive work of Havelock or others on the forms of oral poetry. This may be
another form of the old joke, "Homer didn't compose the Iliad and the
Odyssey; it was another Greek named Homer." Thompson was the director
of the British Library from 1898 to 1909 and is the author of the initial part
of the *Encyclopedia Britannica* entry (otherwise unsigned) cited above.

18. See Leonard, Chap. 3., entitled "Unfeeling Accuracy." He identifies
the 1850s (particularly in the wake of the Lincoln-Douglas debates in Illinois) as the watershed in American political reporting. I concentrate below
on reporting in the federal government, but I do not mean to deny related
versions of the same history in state and local government.

19. Many of these details were given by D. F. Murphy, a congressional
reporter for more than three decades and ultimately chief official reporter
of the Senate. Murphy is quoted in the *Transactions* of the first International Shorthand Congress (1887), 56, 62. He gave the same $50,000 fig-
ure in 1887 that the New York State Stenographers' Association gave for
1879; *Proceedings* (4th annual): 103. For issues of access and democracy,
see Warner, esp. Chap. 3.

20. *The National Stenographer* 1 (1890): 82.

21. In England things happened differently and earlier. Thomas Gurney
was appointed shorthand reporter to the Old Bailey in 1737. In the early
nineteenth century W. B. Gurney was appointed shorthand reporter to both
houses of Parliament. The Gurneys used their own system of brachygraphy,

based on William Mason's *Plume Volante* (1707). For an additional discussion of parliamentary reporting in the United States and United Kingdom, see the introduction to Bickford et al.

22. Edison's reputed trip to Washington is noted in an interview with George Parsons Lathrop. Edison was based in Boston at the time, and there is no evidence of a trip to Washington. On the vote recorder, see Edison, *The Papers of Thomas Edison*, 1:84–85.

23. See Tinling for a good summary of Lloyd's activities and an evaluation of his shorthand system and abilities. I am grateful to the Philadelphia Archdiocesan Historical Research Center and the First Federal Congress Papers Project for suggestions about Lloyd.

24. *Supplement to the Congressional Globe: Containing the Proceedings of the Senate Sitting for the Trial of Andrew Johnson, President of the United States*, 40th Cong., 2d. sess., 1868, 93–94.

25. See Schudson, on the objectivity of the professions, among which he includes law and journalism, but not court reporting. On reporting testimony, see Woodward.

26. T. Anderson, 222; Clephane's testimony from *Supplement to the Congressional Globe: Containing the Proceedings of the Senate Sitting for the Trail of Andrew Johnson, President of the United States*, 40th Cong., 2d sess., 1868, 94.

27. Goody, e.g., "In the jural systems of societies without writing, there can be no effective distinction . . . between law and custom" (*Logic of Writing*, 130). On law reports (judicial decisions) in America, see Surrency, who makes the point that "methods of reporting have not so much progressed as merely changed in accordance with the changing needs of the [legal] profession" (49).

28. Reported testimony was indeterminate in earlier cases where little is known of reporting methods. The Salem witch trials, for instance, were "digested" by clerks. Any discussion of those proceedings must therefore weigh the witch trial (matter) against the witch hunt (method). Similarly, I would propose that Carlo Ginzburg is saying as much about himself as about his sixteenth-century miller when he writes, "It was not the books as such, but the encounter between the printed page and oral culture that formed an explosive mixture in Menocchio's head" (51). Ginzburg is "reconstructing" (xiv) Menocchio from the written records of his oral Inquisition trials and only partially admits the explosiveness of this endeavor.

29. The nineteenth-century ascendance of shorthand in the courts coincided with an ascendance of statute. The full codification of common law desired by some jurists, like David Dudley Field of New York, never succeeded. Shorthand reports may ironically have served Joseph Story's

conservative, anticodification side of the debate by adding so many sure, individual expressions of the common law to the record (Schudson, 96–99, 121); see also Miller's comments and anthology of the "legal mind in America," and particularly selections therein from Field; Walker; and Sedgwick.

30. See Surrency, 58.

31. See McGill, on this case and the logic of copyright; reported as 33 U.S. 591 (1834).

32. See D. Cohen, 26–29 particularly. The closeness (often identity) of trial reports and newspaper reports bears emphasis. Like news, court proceedings have no author, yet they can both arise within a highly competitive marketplace.

33. Titles vary. The Lincoln case was published by Moore, Wilstach & Baldwin in 1865. The Ku Klux case was published in 1872 and reprinted by Negro Universities Press in 1969. Both bibliographic records are according to the Library of Congress cataloguing.

34. Graham was famous (at least in reporting circles) for having reported a celebrated murder trial. To the extent that other court proceedings had been reported and published, they too had possessed the same confusion of authors. Thomas Lloyd published some early American examples of the genre.

35. Kittler characterizes the romantic "discourse network" of 1800 with reference to the "Mother's Mouth," the domestic and feminine scene of literacy acquisition.

36. See Aarsleff, for a basic summary of philology in Britain; Andresen, for the United States; and Christy, on the role of geology. In the material on "vocabularies" below, I am relying on my "The World Recounted."

37. *Transactions of the First International Shorthand Congress*, 181–82.

38. See Baron, 86–88, 92.

39. This account of spelling reform is given in Mencken, 397–407; Baron, Chap. 4; see also Pitman and St. John, Chap. 6. For a lucid, comparative account, see Angel Rama on spelling-reform movements in postcolonial Latin America (43–45).

40. See Mencken, 320.

41. See von Humboldt; and Aarsleff, xii. On the antecedents of the IPA, see Andresen, 98–100. For a sense of the lingering issues that divide phonetic alphabets, see Hollow.

42. Alexander J. Ellis wrote approvingly of Bell's "visible writing" system, which resembled his own phonetic alphabet. The younger Bell was also a student and devotee of "visible writing" and credited his father's system as one inspiration behind telephony. The copy of this document that I

have examined is in the general collections of NYPL. "Regulation" of bodies and affects was perhaps most noticeable when phonography was introduced into elementary schools as a pedagogic tool. In 1852 phonography was used in the schools at Waltham, Massachusetts, and a grandson of Isaac Pitman later wrote, "The fonetic print corrected the brogue of Irish children and the Yankee dialect" (Pitman and St. John, 85).

CHAPTER 2, *Imagining Language Machines*

1. Edison, *The Papers of Thomas A. Edison*, 1:656.
2. See Andresen, 40.
3. Edison, "The Phonograph and Its Future," *North American Review* 126 (1878): 533–34. This article was widely reprinted. Edison's handwriting was swift and legible, its letters small and disconnected. See also Edison, *The Papers of Thomas Edison*, 1:75–76.
4. *Telegraphic Journal and Electrical Review* 7, no. 142 (Jan. 1, 1879): 1.
5. See Hubert.
6. Dowling, esp. Chap. 4, entitled "Disembodied Voices."
7. For very different, yet largely complementary, versions of this cultural change, see Nasaw; Peiss; Kasson; and Lears.
8. Yates, esp. Chap. 3, entitled "Genres of Communication."
9. Of course all genres are accidental in a sense, the unconscious and consensual structures of literary and *sub*literary form, unselfconscious "social contracts" (Jameson, 106). But ballad writers usually know that they are writing (or explicitly not writing) ballads as such. By contrast, the authors of the idea letters may have been writing within their peculiar (sub)genre of the epistle without knowing that any such form existed, though this sort of accident was probably rare, since newspapers and popular discourse of other kinds would have included or noted remarkable examples of the genre.
10. See, for example, Wik on Henry Ford; and McElvaine on FDR. Idea letters were received by both and seem to have long been a feature of fame and power in America. For idea letters to Abraham Lincoln (and Lincoln's own interest in invention), see Holzer, 171–94. For similar and pertinent work on a much smaller sample, see Nord. Nord's sample is problematic, but his work provides a reminder that reading newspaper articles is the countervailing act to writing idea letters, often the immediate stimulus for letter writing, and its source of information, misinformation, and accumulated celebrity around the public figure.
11. See Yates; Chandler; and Beniger. Two recent histories of the computer emphasize the predigital culture of information control in descriptions

of the U.S. Census in the nineteenth century and the early history of International Business Machines: see Shurkin; and Campbell-Kelly and Aspray.

12. The "insiders" had their own textual community, well described by Carolyn Marvin. The phrase *culture of letters* literalizes Richard Brodhead's term. For Brodhead, literary production "addresses and helps call together some particular social grouping, a portion of the whole potential public identified by its readerly interests but by other unifying social interests as well" (5). I am using the terms *insiders* and *outsiders* to avoid "professional" and "amateur" inventors as much as possible; see Gorman and Carlson, 158–59.

13. On the matter of Edison's celebrity and the role of newspaper and other publications in tending the flame, see Wachhorst.

14. For a helpful consideration of these issues in the current history of technology, see Scranton; for an earlier characterization of Scranton's interest and on "the road not taken," see Staudenmaier, 175; on teleology inherent to technology studies, see Collins, 163. In Arnold Pacey's terminology, the letter writers possessed the necessary "objectives," but lacked a "discipline" (14). Admittedly the teleology of the history of technology as a discipline originates from its sources. Not until the international expositions at Jamestown and Seattle (1907 and 1909, respectively), for example, were inventors given a venue to display unproven items, and then it was a quasi-commercial exhibit cobbled together under the title "Bureau of Inventions." The expositions usually included small historical exhibits of "progress" in the form of outdated implements and machines that served to enforce the Whiggishness of the occasion.

15. See Dégh for a helpful discussion of modern media and folklore, particularly magic. Technology was a paradoxically magical nonmagic in each of the idea letters; it was socially connective (at least connecting the author and Edison) and redemptive.

16. All of the idea letters quoted in the text are part of the Edison Document File and Edison General File record groups at the Edison National Historic Site (hereafter ENHS). Each letter has been cited by author's last name and year of composition.

17. This "local" feature is perhaps what distinguishes the writers of idea letters from the more mainstream "technological utopians" described by Howard P. Segal, if, in fact, there is any distinction to be drawn; "Because the technological utopians were not members of an organized movement, it has not been easy to define them" (45).

18. See also Cipolla. For time, see Kern; and Macey.

19. *The Telegraphic Journal and Electrical Review* 7, no. 142 (Jan. 1, 1879): 1.

20. See similarly the Arthur Conan Doyle story "The Voice of Science,"

The Strand 1 (1891): 312–17. For "brazen tablets," see *The Telegraphic Journal and Electrical Review* 7, no. 142 (Jan. 1, 1879): 1. Another hint that phonographs were symbolic subversions of social as well as textual order was the idea of a cursing phonograph, apparently proposed by Alexander Graham Bell as early as 1878 (Koenigsberg, *Patent*, 43) and "invented" by Mark Twain's bilious avatar of technological progress, Colonel Sellers, who was created in the novel *The Gilded Age* (1873), which was coauthored by Charles Dudley Warner and later adapted as a wildly successful dramatization. Sellers returns in the less successful play *Colonel Sellers as Scientist* (1883), coauthored by William Dean Howells. Twain later reworked this play into the farcical novel *The American Claimant* (1892). When the second play was produced in New York, Twain crowed in a letter to Howells that Thomas Edison had been approached and would provide a phonograph and electronic marvels for the stage, a veritable "'invention-museum'" that Edison reportedly claimed "'shall be memorable in the annals of the stage.'" He loaned "a phonograph—a telephone—a 'Shocking machine' (large glass wheel with brass attachments) and a dozen or more other odd looking things" (Smith and Gibson, 561, 592). E. H. Johnson, a friend of Edison's at the Edison Electric Illuminating Company, tried to get Edison into the city to see the play, in which, Johnson says, there is a character who "invents everything" (Edison, *Thomas A. Edison Papers: A Selective Microfilm Edition*, 119:225). Knoper is particularly good on Twain's performative aspects. On "Sexual Identity, Death, and the Family Piano in the Nineteenth Century," see Leppert, 119–51.

21. Edison draft of Oct. 8, 1888; Edison, *Thomas A. Edison Papers: A Selective Microfilm Edition*, 113:238. For the microphone controversy in the popular press, see ibid., 25:234–45.

22. *Telegraphic Journal and Electrical Review* 7, no. 142 (Jan. 1, 1897): 1.

23. Sociologists, folklorists, and anthropologists are only now beginning to canvas Western culture for the roots of rumor and misinformation within media-rich society; see Turner, *I Heard It through the Grapevine*.

24. See Foucault. Idea letters like these are suggestive representations of insanity that seem to coincide with Foucault's account of nineteenth-century madness, internalizing surveillance, and perpetual judgment, focused on the Father, the redeemer, the doctor—Edison. Yet by themselves these few letters are ahistorical. Without further knowledge of the social and biographical conditions under which these documents were written, they remain merely suggestive, proving in their isolation, as Foucault says of the relevant archaeology of knowledge, "This is doubtless an uncomfortable region" (ix), where "terminal truths" and diagnoses are questioned as features of broader, discursive operations of society.

25. See Crary; a good compliment to Crary is Brodhead's Chap. 2, entitled "Veiled Ladies: Toward a History of Antebellum Entertainment."

26. Cheape, 6–7; see also Table 4 in Bureau of the Census, *Special Reports, Street and Electric Railways, 1902* (Washington, D.C.: Government Printing Office, 1905).

27. This is from Dreiser's "From New York to Boston by Trolley," reprinted in Hakutani, 2:91–100.

28. *Fourteenth Census of the United States Taken in the Year 1920. Volume I. Population 1920* (Washington, D.C.: Government Printing Office, 1921).

29. Particularly helpful on the American culture of reading is the work of Zboray. On the effect of illumination, see Stewart, *On Longing*, 9; on the fountain pen, Strasser, 109; and on Russia, Coopersmith, 48.

30. No title, *Harper's Magazine*, Aug. 1992, 19–20. The most popular suggestion in 1996 was reportedly to tie up the boulder and drag it into to sea with ships! (Associated Press reports, Feb. 14, 1996).

CHAPTER 3, *Patent Instrument and Reading Machine*

Portions of the chapter have appeared previously as "Reading Music, Reading Records, Reading Race: Musical Copyright and the U.S. Copyright Act of 1909," *The Musical Quarterly* 81 (1997): 265–90.

1. On "textual systems," see Bazerman, "Electrical Connections"; on patents, see Myers; and also Bowker; on wanting to look old and new at the same time, see Bowker, 62.

2. See Edison drafts of Feb. 17, 1888, and Aug. 7, 1897; Edison, *Thomas A. Edison Papers: A Selective Microfilm Edition*, 114:308, 835. Macomber, 14. The legal authorities used in this chapter (therefore my present-tense readings of them) were relevant to the period in question, but they are not reliable sources for intellectual property law today.

3. Patent Act of 1790. Patent acts and statutes are quoted from Walker.

4. The relations of technical knowledge and writing are in particularly sharp focus along the science/technology divide, since scholars have long distinguished science from technology in part by identifying the differing relations each has with discursive networks (see Cutcliffe and Post). To some, science and technology are mirror-image twins (see Layton, "Mirror-Image Twins"): the scientist publishes new findings as soon as possible, making public knowledge. In contrast, the technologist hides new knowledge, protecting and keeping it private. The successful scientific article is supposed to encourage reproducibility (but see Collins, 55, 130); the successful patent document only offers reproducibility as it asserts monopoly. Further dis-

tinctions between science and technology traditionally rest on the assumption that while science is smart, technology is "only" applied science. Albert Einstein the patent clerk was the quaint impediment to Albert Einstein the physicist.

5. See MacLeod; other articles that appear in this same special "Patents" issue of *Technology and Culture* 32 (1991) with MacLeod's work are also helpful.

6. See Bowker, 53; Myers, 92.

7. See Tibbetts. Other essays in the same collection (edited by Lynch and Woolgar) are similarly helpful.

8. See also Ferguson's *Engineering and the Mind's Eye.*

9. 167 F. 977 (1909).

10. 114 F. 926 (1902). See also 214 F. 787 (1914), which gives a summary of the application and litigation history of Edison's claims. Patent no. 589,168 (1897) was reissued as no. 12,037 (1902) and then as no. 13,329 (1911).

11. See Israel and Rosenberg. An invention can be "reduced to practice," as the patent office puts it, only after it is invented. Because a true invention can always be reduced to practice, the inventor doesn't always have to reduce it, although any reduction to practice weighs in an inventor's favor as a way of proving the chronology and accomplishment of invention.

12. Patent no. 640,208, granted 1900. Allen Koenigsberg and Aaron Cramer were both generous with their knowledge of Lambert

13. *The Adventures of Huckleberry Finn* and *A Connecticut Yankee* were sold this way; *The American Claimant* was not; the Edison phonograph frequently was. Regional subscription sales were used for all manner of goods and had been used in the book trade at least since the seventeenth century. Subscription sales of books have survived in the different guise of mail order, book or product-of-the-month clubs—yet another articulation of the distance between producer and consumer. See Feather, 62–63, for a word on the different logic (and law) of price-fixing in the British book trade.

14. *Bobbs-Merrill Co. v. Straus*, 210 U.S. 339 (1908); *Straus v. American Publisher's Association*, 231 U.S. 222 (1913); and *Straus v. Victor Talking Machine Co.*, 243 U.S. 490, 494–95 (1917). *Victor* was a final affirmation of the Court's decision in "the Santogen case," *Bauer v. O'Donnell*, 229 U.S. 1, 8–9 (1913). See Vaughan, 127–33; also Hower, 352–57.

15. There is an echo here of nineteenth-century debates surrounding paper money. See O'Malley. The essentialism that he finds coincidental in economic and racial discourse of the late century is exploded in different ways by the technology of recorded sound. See below.

16. 213 U.S. 325 (1909); and 152 U.S. 425, 432–33 (1894), respectively. See Vaughan, 173–74, 178.

17. 243 U.S. 502, 510–11; and 243 U.S. 490, 494–95, respectively.

18. Vaughan, 127–33, 174–77; on use-values, Attali, 96.

19. Alexandra Mullen pointed me toward many such anecdotes; these are from Grimsted, 60. See also L. W. Levine, *Highbrow/Lowbrow*, 30. On a related blurring, see Davidson on the real grave of fictional Charlotte Temple.

20. See Carolyn Marvin on the anecdotes and constructed expertise of telegraphy and engineering publications. Of course, popular culture fed and reinforced the same exclusionary rhetoric; recall the spectacularly racist second verse to Stephen Foster's "Oh! Susanna" (1848), performed by the African American troupe, Christy's Minstrels: "I jumped aboard de telegr[a]ph / And trabbelled down de river, / De Lectric fluid magnified / And Killed five Hundred Nigger" (*Democratic Souvenirs*, 104).

21. On the ethnographic present, see Fabian, *Time and the Other*; and Clifford and Marcus. Examples of the type-scene appear in the *Music Trade Review* (Dec. 21, 1907), in Robert Flaherty's *Nanook of the North* (1922), and elsewhere. For a discussion of type-scenes as such, see Pratt. See also Peitz, for a poststructuralist meditation on the phonograph and its colonialist relations. The whole, varied history of ethnicity and the phonograph has yet to be written; I am grateful to Jerry Fabris for sharing his thoughts on the phonograph and ethnomusicology.

22. See Laing. "Love & Theft" is Eric Lott's title; my debt to Lott on minstrelsy will be clear below.

23. This cite refers to *Arguments Before the Committees on Patents . . . December 7, 8, 10 and 11; 59th Congress*. Brylawski and Goldman's work will be cited by page number in the text below; the different sets of hearings are paginated individually in Volume 4.

24. Like early motion pictures, phonograph records partook of the culture of vaudeville, thriving on the same heterogeneity of participants and a similar variety of acts, rooted in the dynamics of performance and a tumult of oral forms. Edison's National Phonograph established its New York offices on Union Square, in the heart of the vaudeville district. See Snyder, 58–59, on the class and racial heterogeneity of the vaudeville; and on variety experience, see Snyder, 106. For a corporate genealogy of Columbia, RCA, and other companies, see Read and Welch.

25. *Burrow-Giles Lithographic Co. v. Sarony*, 111 U.S. 53 (1884).

26. Hearing on June 6; see "laughter," Brylawski and Goldman, 165. Attendance was all male; Oscar Wilde was by then disgraced and dead.

27. See also Edelman, Pt. 2. The analogy between music and photo-

graphs came up at the hearings in part because the complications of locating authorship *and* subjecthood in a copyrightable form. American copyright law has returned repeatedly (and never with great satisfaction) to issues of personality.

28. 122 F. 240 (1902); Edison won on appeal in 1903, the same year another important precedent was set in one of the *Edison v. American Mutoscope and Biograph Co.* cases; see Musser, *Before the Nickelodeon*, 238.

29. See registration form for Thomas Crahan's "Artistic Glimpses of the Wonder World" (1900), Edison Document File copy, ENHS. On the matter of deposits, see "Copyright" folders, 1908–1910, Edison Document File record group, ENHS; as well as Loughney, 60 and following. The number of deposits in 1905 was given by the librarian of Congress, Herbert Putnam, during his statement in *Arguments before the Committees on Patents* (Brylawski and Goldman, 14).

30. The U.S. Census reported in 1905 that the annual value of American sheet music had risen from $1.68 million in 1890 to $2.27 million in 1900 and to $4.15 million in 1905. In the same scheme, American pianos produced for 1905 were valued at $69.6 million, while the annual value of phonographs and supplies had surged to $10.2 million in their two decades on the market. These figures were repeated into the record of the congressional copyright hearings as evidence of the vigor that different segments of the music industry possessed.

31. This from an Edison record catalogue for blue-amberol record no. 2147, "On the Banks of the Wabash, Far Away," performed by baritone and chorus with orchestra.

32. See Dreiser's "Birth and Growth of a Popular Song," *Metropolitan* 8 (1898): 497–502, reprinted in Hakutani, 2:19–22. See also Dreiser.

33. See Sanjek, Chap. 11. Middleton's methodological reflections on popular music were helpful to these several pages, particularly regarding the varied meaning of "popular" in studies of popular music; of related interest is Lewis's article on popular music as "symbolic communication."

34. Edison undated memoranda, covered by Frank L. Dyer letter of Feb. 26, 1908; Edison Document File, ENHS. For a cultural history of Sousa as well as Sousa and "The Culture of Reassurance," see Harris.

35. 209 U.S. 1 (1908); see also 147 F. 226 (1902).

36. See Laing, 7–8.

37. See Lott. In the postminstrelsy age, popular music in general seems to fulfill this role in another way; I am thinking of Paul Gilroy's sensitive polemic, Chap. 3, entitled "'Jewels Brought from Bondage': Black Music and the Politics of Authenticity."

38. On contemporary questions of essentialism, I have been particularly

influenced by Michael O'Malley and Nell Irvin Painter's succinct contributions to the *American Historical Review* Forum on race, money, and "intrinsic value" in nineteenth-century America. O'Malley wrote, "Facing the possibility that [white-skinned "black"] men such as Plessy could renegotiate racial value . . . the court responded with irrational theories of intrinsic racial difference" (395).

39. See Dorman; see also Sanjek, Chap. 9.

40. According to Riis, in the period "just before jazz," Black music "seems to have meant syncopated tunes or dialect songs on a nostalgic, Old South theme" (154).

41. *Edison Phonograph Monthly* 3, no. 5 (July 1905): 10.

42. See Lott, 20.

43. See Turner, *Ceramic Uncles & Celluloid Mammies*, 20, 22, e.g., for an observation of the differing modes of aural and visual racism in popular culture.

44. See Sanjek, 297.

45. Here I am indebted to conversations with Miranda Paton regarding her in-progress work on the iconography of the phonograph and to an e-mail exchange with David W. Stowe about what he calls "racial ventriloquism." Interestingly enough, phonographs and records remain tenaciously visual artifacts for collectors; see Schwartzman. The "Whole World Kin" advertisement appeared widely; copy in Primary Printed collection, ENHS.

46. Edison's phrenological notes are in "Employment" folders, Edison Document File, ENHS. James and Edison met on a steamer for Europe; Edel and Powers, 329.

47. "Perfected Phonograph," *North American Review* (1888): 641–50, quote on 645. Also see Irwin's extended analysis of the hieroglyph metaphor. In one technical note from November 1877, before the phonograph had been realized, Edison even dilated on its musical potential in textual terms: "Reproduce from [tinfoil] sheets music both orchestral instrumental & vocal the idea being to use a plat machine with perfect registration & stamp the music out in a press from a die or punch previously prepared by cutting in steel or form an electrotype or cast from the original on tin foil" (Edison, *The Papers of Thomas Edison*, 3:629; punctuation added).

48. Levin, 56. Levin's "For the Record" includes translations of Adorno's "The Curves of the Needle," "The Form of the Phonograph Record" (quoted here), and "Opera and the Long-Playing Record." The imagined universality of (usually Western) music jibed with other attempts at universal language. Esperanto (1887) had a grand launch during the Paris exposition of 1900 (Mandell, 68), and there were other stabs at the same thing: IDO (described by L. De Beaufront); the Master Language (described

by Stephen Chase Houghton); Word-English (by Alexander Melville Bell); and Tutonish (Elias Molee). All examples come from NYPL, General Collections. See also Eco, esp. Chap. 16., on international auxillary languages.

49. See Brylawski and Goldman, 106.

50. Ibid., 157.

51. Some European developments are documented in the records of the National Phonograph Company, ENHS. The British case, *Neumark v. National Phonograph Ltd.*, is described in a letter from G. Croyden Marks to William E. Gilmore, Apr. 10, 1907, Legal Department Records, ENHS. On France, see Attali, 97–98.

52. Letter from C. A. L. Massie to H. C. Kennedy, July 19, 1907, regarding *Jose Elizondo v. Jorge Alcade*, located in Legal Department Records, ENHS. The copyright in question was a selection from a comic opera entitled "El Chin Chun Chan."

53. See Read and Welch, 393.

54. 175 F. 875 (1910). See Kaplan, 41; on Hand, see Cracas.

55. See L. Cohen, 105.

CHAPTER 4, *Paperwork and Performance*

1. See Garbit, 7; Edison, *The Papers of Thomas Edison*, 3:656, notebook entry by Charles Batchelor; and Atkinson's edition of Ganot's *Physics*.

2. Two qualifications: electric lighting may be a bad contrast in this illustration, since, as Carolyn Marvin argues, electricity shared some of the features of communications media. Electric light is McLuhan's medium as/without message (8). Similarly, the biological bases of language acquisition may be too aggressive a comparison: it should be clear that I am *not* proposing that media fully make our experience, that recorded sound itself constructed the ears that heard it. I'd like to thank Dave Heitz for demonstrating his replica of the tinfoil phonograph, and Jerry Fabris of the National Park Service for his demonstrations of later machines.

3. This "tailored" material culture is clearly related to Banta's "taylored" lives, though I don't want to fudge the chronology or the influence of scientific management as it was later conceived. My thinking about inscriptions in the laboratory and the marketplace has been influenced most by the work of sociologists of science; see Latour, "Drawing Things Together." See also Latour's *Science in Action*; and Latour and Woolgar's *Laboratory Life* for amplifications of this perspective.

4. Doheny-Farina (3, 30) is my source for technology transfer. *Technological imperative* is a necessary term I've adopted from Hoke, in particular.

5. See Cooper for a good example of product revision. If producer-

narratives can be wrong, then so can their revisions. The danger is learning nothing about consumers. Only when combined with some sense of "success," of consumer practice, can either narrative or revision tell us about consumers. I am sympathetically wary of gestures like this one in a history of advertising: "Nor have I asked whether or not a particular advertising campaign has helped to sell a particular product. This question does not, in my judgement, reveal much about the broader cultural significance of advertising" (Lears, 3), since cultural significance so richly and reciprocally encompasses the local and material.

6. See Orvell, xvii–xx.

7. Undated item, 1902 Motion Picture, Edison Document File, ENHS.

8. See F. K. Dolbeer memorandum, Aug. 6, 1908, and accompanying correspondence, especially Frank L. Dyer to Edison, Aug. 12, 1908, records of the National Phonograph Company, ENHS.

9. Records of the National Phonograph Company, ENHS.

10. 14 F. 728 (1883). The Mark Twain Case was heard amid the context of another federal case of related import, "The Sarony Case" (see above); see Gaines, 81.

11. See Redmond, 8, 11, for examples of works falsely attributed to accomplished authors.

12. U.S. trademark no. 34,806.

13. See "EUREKA!" and following, Browning et al., 441–42.

14. Shakespeare was not the test inscription for the phonograph. Kittler makes much of the fact that Edison's first record was of a childhood rhyme, "Mary Had a Little Lamb," *Discourse*, 231–32; see below. Paul Israel tells me that *Richard III* was the young Edison's favorite.

15. See Edison, *The Papers of Thomas Edison*, 2:483–579, on the electric pen.

16. Letter to the Editor, Sept. 29, 1878, *Chemical News* (Oct. 18, 1878): 198.

17. Edison marginalia with triple underlining, Feb. 17, 1909, minutes of the meeting of the Executive Committee of the Edison Manufacturing Company, Edison Document File, ENHS.

18. For the chronological development of Edison record labels, see Koenigsberg, *Edison*, 152–58.

19. For an invaluable analysis of the economy of "stars," see Dyer.

20. See Strasser, 31–35. Related manifestations of a similar impulse were the vaudeville and movie "palaces," the architectural packaging of the theater and motion picture experience. For the Essex Press scandal, see hints in the minutes of the National Phonograph Company, Edison, *Thomas A. Edison Papers: A Selective Microfilm Edition*, 156:297–440.

21. Koenigsberg, *Patent*, 52, 56. See also Dethlefson.

22. I do not mean to imply a causal relation between disc surface area and the ultimate success of the disc. The "success" of the disc record and concomitant "failure" of the cylinder, a market adjustment that unfolded during more than thirty years, has rightly been ascribed to many causes, the least convincing of which are purely technological. For one pertinent narrative, see Siefert on "How the talking machine became a musical instrument"; Siefert rightly attends to naming and "published discourse" in her account of how Victor succeeded in commodifying a new standard of fidelity that was more "Cultural" than technological.

23. U.S. patent no. 713,863.

24. On disc labels and picture records, see Koenigsberg, *Patent*, 56, 59; on insert slips, see his *Edison*, 151.

25. See Dethlefson, 26.

26. Ibid., 10–11.

27. The map is Jorge Luis Borges's figure, cited in Baudrillard.

28. A. N. Petit in N-09-01-04, ENHS.

29. By "mark-up language," I am referring not only to languages like XML, SGML, and HTML but also to projects like TEI (Text Encoding Initiative) or EAD (Encoded Archival Description), which seek to mark formal textual components with such languages.

30. Musser, *Before the Nickelodeon*, 39 and following. Look for a restored version of this film clip in the film version of Vito Russo's *Celluloid Closet*. Charles Musser tells me that the violin player cannot be identified as Dickson with absolute certainty.

31. See Hounshell.

32. Affidavit filed in *Edison United Phonograph Company v. Thomas A. Edison et al.*, June 17, 1895, Chancery Court of New Jersey, copy at ENHS.

33. Edison Document File record group, ENHS.

34. Patents no. 470,477, granted Mar. 8, 1892, and no. 754,825, granted 1904. The Criswells' repeating raven was intended as an advertisement for their patent medicine (a remedy for corns). Spencer and Lynde were both from Atlantic City, and their "speaking figure" may have had a similar advertising purpose.

35. See George Croyden Marks to Frank L. Dyer, Apr. 10, 1907, Legal Department Records, ENHS

36. This alienation of the performer is another one of Walter Benjamin's points in his celebrated "Work of Art in the Age of Mechanical Reproduction."

37. 214 F. 787.

38. See Frank, 3, on this context and, more particularly, on Villiers's "foregrounding of the phonograph as instrument of technological and social transformation" (143).

39. This is the interpretation of Read and Welch, which has not been sufficiently explored or contested by more recent scholarship.

40. See Meikle, 8, 189.

41. My sense is that discs may have been pirated less than cylinders, which had less to do with technological differences than economic ones. Siefert, Thompson, and others have explored the issue of musical fidelity, but neither they nor I explore what appears to have been the slow-in-coming realization that records, not phonograph machines, (software, not hardware) could drive the industry. Records were not just secondary supplies, like vacuum bags or mimeograph paper, sold as a consequence of another purchase. Because they were cultural, they were somehow primary products in their own right. European owners of Edison's rights, and later his branch offices and sales companies (pace Siefert), all would likely have had more success had they realized this. On "distress," see Stewart, *Crimes*, 66–101; the similarity and the distance between literary genres and cultural artifacts like phonograph records are both evocative in this context.

42. See Thompson on the tone test, for example.

CHAPTER 5, *Automatic Writing*

1. For Batchelor's comment, see Edison, *The Papers of Thomas Edison*, 3:699. "Repertoire" and "style" are from Jenkins and were helpfully discussed in many conversations I have had with Terry Collins, Gregory Field, and Paul Israel.

2. See Leary. Like comparative philology and ethnomusicology, the new cognitive psychology mixed imported, German thought with American academic and other cultural conditions. Notably the German pioneer Wilhelm Wundt trained American students like James McKeen Cattell, an early experimenter in the area of cognition and reading research. Automatic writing seems a particularly American interest in psychology, unlike reading, aphasia, and the other elements of "psychophysics" (Kittler, *Discourse*, esp. 222, 225).

3. *Psychological Review* 2: 329; emphasis in original.

4. Readers familiar with Kittler will recognize the similarity between our interests. Kittler's central gesture of disjunction or rupture, between the discourse network he calls "1800" and the one he calls "1900" (with the latter's identity of typewriter, film, and gramophone with the Lacanian orders of the symbolic, imaginary, and real), as well as his neglect of the (per-

haps uniquely American) spiritualist tradition and the context of telegraphy and other business communications, tend to make this account of the typewriter itself more symbolic than it is "historically concrete" (pace Wellbery, in Kittler, *Discourse*, xii, xxx). On Nietzsche, see also Stingelin.

5. Originally from "What Psychical Research Has Accomplished," *Scribner's* (Mar. 1890).

6. A good primer on gender and technology is Wajcman; on the sexual division of labor, see 29–42, including typesetting and clerical work. Also see Kwoleck-Folland, which offers a sophisticated and enlightening analysis of the gender relations of the modern office.

7. The census of 1890 was the first for which "Stenographers and typewriters" was a separate occupational category. By then there were 33,418, 63 percent of whom were women. In 1910 the number of people employed as stenographers and typists had increased nearly tenfold, and 83 percent were women, or more than a quarter of a million individuals. The lack of comparable census data for 1880 is frustrating, for then there were a half a million people counted as employed as clerks, copyists, bookkeepers, and accountants, categories into which stenographers and typists disappeared. Only 6 percent of this half million were women.

8. Letter of Jan. 31, 1911; Bates Manufacturing Company record group, ENHS. On "mechanical objectivity," see Porter. On the word *automatic* and its gradual semantic encroachment on *manufacture*, see Pacey, 268–69.

9. Automatic machines of the last century encapsulate managerial bureaucracy in much the same way that cyborgs are now inveighed to encapsulate an activist politics. Given currency in Donna Harraway's socialist, feminist "A Manifesto for Cyborgs," the cyborg involves the celebration of an antiessentialist collapse of dichotomies between organic and mechanical, woman and machine. One of the reasons the cyborg works so well as a figure for Harraway is its critical disjuncture from an older logic of automata, including narratives of bureaucracy, management, and the experiments in social control that Banta calls "taylored lives." Goldberg (11–12) suggests that Harraway's perspective be brought to bear more explicitly on accounts of writing.

10. Leo Marx's *Machine in the Garden* remains one of the most convincing accounts of how the railroad mattered in American culture, particularly to canonical authors and American landscape artists. For accounts of the telegraph, see Blondheim; Clayton; as well as Marvin. The distinction between railroads and telegraphs might be seen as one of "technological sublimity." *Technological Sublime* is Marx's term, appropriated by Nye for the title of his recent work.

11. Craig qtd. in Edison, *The Papers of Thomas Edison*, 1:246; Edison, *Thomas A. Edison Papers: A Selective Microfilm Edition*, 12:447; emphasis in original. The earlier language of Mesmerism offers a contrast equally suggestive of renegotiated subjectivity: like the telegraph and then the telephone, Mesmerism relied upon an individual called the "operator," the hypnotist, who served doubly as agent and intermediary.

12. Craig qtd. in Edison, *The Papers of Thomas Edison*, 1:251. See Current, 42–43. The machine Sholes showed to Edison was his "axle" machine, which wrote around the cylindrical platen instead of along it; that is, it had the same axis of inscription that Edison's phonograph (not his typewriter) later would. Just as Edison was perfecting his phonograph, another typewriter inventor and entrepreneur, James Hammond, approached him, seeking help with his prototype, a typewheel machine; Edison, *The Papers of Thomas Edison*, 3:548; Edison, *Thomas A. Edison Papers: A Selective Microfilm Edition*, 14:158.

13. See Knoper, 126.

14. See Barborka.

15. I am relying on Koutstaal here. Admittedly, even nonautomatic writing was a favorite sign of either normality or pathology during the period as psychologists such as Martin Charcot studied "linguistic disorder[s] caused by lesions in specific parts of the brain, scars that were themselves described as 'indelible' writings on the mind" (Panchasi, 13). For Freud's dependence upon writing metaphors, see Derrida. Freud's early interest (exhibited in *The Psychopathology of Everyday Life*, for instance) in involuntary speech acts, that is, slips of the tongue, makes an important, if perhaps incomplete, distinction between his work (the "talking cure"; the mind as writing machine) and the contemporary work of psychologists who worked on automatic writing. Kittler's account of psycho-physics and its stakes is particularly rich ("The Great Lalulä," *Discourse*).

16. See Collins, 30–33.

17. Mrs. Underwood contributed a previous article on the phenomenon; Mr. Underwood contributed this and a sequel. There is no relation, that I know of, between these Underwoods and the Underwood typewriter.

18. On spiritualism and American literature, see Kerr; and also Strout, 40–51. On Nathaniel Hawthorne's spiritualism, see Brodhead's Chapter 2, entitled "Veiled Ladies." The subject of spiritualism interestingly appealed across a supposed spectrum of styles, from symbolists and romantics like Hawthorne and Poe to "realists" such as Henry James and Howells.

19. See Garland, *Forty Years*, 113, "The messages were again disappointing. They had no relation to me, none whatever"; that Garland's questions were typical, see Isaacs, 90. In his other memoirs, like *Son of the Mid-*

dle Border, Garland's partial representations of his relationships with women are almost as odd.

20. *New York Times*, Dec. 1, 1910, 11.

21. Epinetus Webster, *Phonographic Teacher* (1852): (xii).

22. Edward Easton of the American Graphophone Co., Rowland F. Cromelin of the Columbia Phonograph Co., and James Clephane were all former court reporters, as were many of the early backers and inventors of typewriting machines. On the typewriter, see Yates, 41–43; and Davies, 31–38; both rely heavily upon works by Current.

23. Garland boasted, "Edison once said to a friend of mine, 'All along my way I've come upon hints of these mysterious forces—and sometime I am going to stop commercial inventing and follow out the these leadings'" (*Forty Years*, 134). Crookes was the inventor of the Crookes's tube, which was vital to Roentgen's discovery of X rays, and was a devotee of Theosophy and psychical research. Another scientist who dabbled in spiritualism was Alfred Russell Wallace.

24. Reprinted in Twain, 224.

25. See Currant, 72; Twain's typing was all in capitals. Charles Dodgson was another pseudonymous author (Lewis Carroll) who purchased an early typewriter, in Dodgson's case a Hammond, with its semicircular keyboard. Like Twain, Dodgson experimented with many mechanical writing aids. Even more than Twain, he saw the typewriter as a mere novelty and put it to limited use, mostly to impress little girl correspondents; see Lovett.

26. I would fault Yates; Davies; and Kittler. Kittler declares categorically, "Before the introduction of John T. Underwood's 'view typewriter' in 1898 [*sic*], all models (much to the disadvantage of their popularization) wrote invisible lines, which became visible only after the fact" (*Discourse*, 195). This misperception has entered public history too: *L'HOMME MACHINE*, an exhibit of the Conservatoire National des Artes et Metiers, Paris, humorously provided in the English translation: "It took fifty years before the typist could see what he or she was typing" (Sept. 1994). See Mares, 183. See also Masi, 88. The Underwood was not as revolutionary as has been described. Many previous typewriters had what was ultimately called "visible" writing. Most, like Edison's early typewriter and many of his printing telegraphs, were slower, typewheel machines: their type was located on a wheel that rotated to the appropriate position when a key was first depressed, or on a wheel that the operator rotated by hand; then the whole wheel was pushed against the paper (or the paper against the wheel) when the key was fully depressed. Two front-strike (typebar rather than typewheel) machines, called the Prouty and the Daugherty, also predated the Underwood. Nor did the "visible" Underwood immediately and totally capture

the typewriter market, as a number of accounts imply. The Remington Company reportedly sold 400,000 of its "invisible" machines between 1894 and 1914, after the Underwood was introduced (Underwood sold 500,000 machines between 1894 and 1911), according to Masi.

27. Lucy C. Bull, "Being a Typewriter," *Atlantic Monthly* 76 (Dec. 1895): 827.

28. See Masi, 111.

29. Typewriters, Tenner writes, were early "on their way to being black boxes in the technological sense: mechanisms opaque to the user" (162). Maps of keyboards seem odd, alphabetic holdovers from shorthand manuals, spellers, and other instructional texts, since few students must have studied typing without access to a typewriter. As Kittler notices so brilliantly, the keyboard (also keyboard maps) gives the alphabet space as well as series (*Discourse*, 193–95). Letters themselves relate in spatial coherence that is prelexical, alphabetic, as they would have in earlier printing and optical telegraphs.

30. This analogy between printing press and typewriter is not meant to deny the important distinction of publication that exists between them. The press offers "publicity," while the typewriter offers a complicated renegotiation of private words. Typing must have been empowering for women in several respects. It not only provided an eventual white-collar labor niche (as opposed to the industrial and needle trades) but also women authors could produce typescripts (as opposed to the "scribbling" long dismissed by the publishing establishment), and women secretaries got to know their employers' private business (an empowerment not without further dimensions of risk and sexual politics).

31. See Purcell, 64, 70.

32. Purcell gives an account of the novel's publication and success but strangely misreads the plot of the novel (63). Leslie Fishbein suggested *Trilby* to me and helped with sources on the novel.

33. See O'Brien; see also McLuhan, 258–64.

34. *Self-forgery* was one of the more bizarre concepts of the French graphologist Bertillon, who testified in the Dreyfus case that Alfred Dreyfus has penned an incriminating letter, but had cunningly disguised his handwriting to look like an imitation of his own. Largely on the grounds of this handwritten document, Dreyfus was convicted in 1894. See Panchasi's wonderful account of the graphical mess; see also Brendin. The Dreyfus case attracted worldwide attention and adds depth to the contemporary appetite for anti-Semitic *Trilby*, with its Parisian setting.

35. Primary Printed collection, ENHS.

36. Arthur Conan Doyle (1859–1930) was incidentally a dabbler in

psychical research and spiritualism. Ron Thomas suggested this Sherlock Holmes tale to me.

37. Kwoleck-Folland offers an important reminder that "Gender relations became an important aspect of office work not because of any inherent qualities of men and women, nor simply because of women's presence. Rather, office work encoded ideas about gender in complex new ways" (41).

38. "The History of Touch Typing" pamphlet, Remington Typewriter Co. [n.d.], NYPL.

39. See Edel and Powers, xxiii.

40. See Bolter, for use of the phrase "writing space" in his book about the computer, hypertext, and word processing.

CODA

1. See Shurkin; Campbell-Kelly and Aspray; and R. Rosenberg, 52–61.

2. See, e.g., Landow, *Hypertext*, 10–17.

3. See Duguid, 85.

4. See Landow, *Hypertext*, 27–28.

5. See Lanham, 9.

6. For the history of literature as an academic subject, see Graff. Paulson gives a thoughtful account of relevant interdisciplinary pressures and recent trends (Chap. 1); Lanham connects the recent debates over the Western canon to his accounts of digital textuality.

7. See Czitrom, 73–74.

8. Controversy over the reliability of wirelessness centered around the ill-equipped and understaffed ships that might have come to the rescue of *S.S. Titanic* passengers in time; see Douglas, 226–33. Controversy over the reliability of the Internet involved the circulation of questionable evidence that TWA Flight 800 was shot by an American missile.

9. See Hunt.

Bibliography

Aarsleff, Hans. *The Study of Language in England, 1780–1860.* Princeton, N.J.: Princeton University Press, 1967.

Anderson, Benedict. *Imagined Communities: Reflections on the Origin and Spread of Nationalism.* Rev. ed. London: Verso, 1991.

Anderson, Thomas. *History of Shorthand with a Review of the Present Condition and Prospects in Europe and America.* London: W. H. Allen, 1882.

Andresen, Julie Tetel. *Linguistics in America, 1769–1924.* London: Routledge, 1990.

Atkinson, E. *Elementary Treatise on Physics, Experimental and Applied. Translated and Edited from Ganot's* Elements de Physique. 9th ed. New York: n.p., 1879.

Attali, Jacques. *Noise: The Political Economy of Music.* Trans. Brian Massumi. Minneapolis: University of Minnesota Press, 1985.

Baker, Arthur M. *How to Succeed as a Stenographer or Typewriter: Quiet Hints and Gentle Advice by One Who "Has Been There."* New York: n.p., 1888.

Baldwin, Neil. *Edison: Inventing the Century.* New York: Hyperion, 1995.

Banta, Martha. *Taylored Lives: Narrative Productions in the Age of Taylor, Veblen, and Ford.* Chicago: University of Chicago Press, 1993.

Barborka, Geoffrey A. *H. P. Blavatsky, Tibet, and Tulku.* India: The Theosophical Publishing House, 1966.

Baron, Dennis E. *Grammar and Good Taste: Reforming the American Language.* New Haven, Conn.: Yale University Press, 1982.

Barthes, Roland. "Writing the Event." In *The Rustle of Language,* by Roland Barthes; trans. Richard Howard, 149–54. New York: Hill and Wang, 1986.

Baudrillard, Jean. *Simulations.* Trans. Paul Foss, Paul Patton, and Philip Beitchman. New York: Semiotext(e), 1983.

Bazerman, Charles. "Electrical Connections: Letters to Thomas Edison in Response to His Claim of Solving Incandescent Lighting, 1878." *Journal of Business and Technical Communication* 8 (1994): 135–47.

———. *Shaping Written Knowledge: The Genre and Activity of the Experimental Article in Science.* Madison: University of Wisconsin Press, 1988.

Bell, Alexander Melville. *Universal Line-Writing and Steno-Phonography; On the Basis of "Visible Speech."* London: n.p., 1869.

Bellamy, Edward. "With Eyes Shut." *Harper's New Monthly Magazine* 79 (1889): 736–45.

Beniger, James R. *The Control Revolution: Technological and Economic Origins of the Information Society.* Cambridge: Harvard University Press, 1986.

Benjamin, Walter. *Illuminations.* Ed. Hannah Arendt; trans. Harry Zohn. New York: Schocken, 1968.

Bickford, Charlene Bangs et al., eds. *Debates in the House of Representatives. First Session: April–May 1789.* Baltimore: Johns Hopkins University Press, 1992.

Bijker, Wiebe E. *Of Bicycles, Bakelites, and Bulbs: Toward a Theory of Sociotechnical Change.* Cambridge, Mass.: MIT Press, 1995.

Blondheim, Menahem. "When Bad Things Happen to Good Technologies: Three Phases in the Diffusion and Perception of American Telegraphy." In *Technology, Pessimism, and Postmodernism,* ed. Yaron Ezrahi, Everett Mendelsohn, and Howard Segal. Amherst: University of Massachusetts Press, 1994.

Bolter, David Jay. "Ekphrasis, Virtual Reality, and the Future of Writing." In *The Future of the Book,* ed. Geoffrey Nunberg. Berkeley and Los Angeles: University of California Press, 1996.

———. *Writing Space: The Computer, Hypertext, and the History of Writing.* Hillsdale, N.J.: Lawrence Erlbaum Associates, 1991.

Bolter, David Jay, and Richard Grusin. "Remediation." *Configurations* 4 (1996): 311–58.

Bowker, Geof. "What's in a Patent?" In *Shaping Technology/Building*

Society: Studies in Sociotechnical Change, ed. Wiebe E. Bijker and John Law. Cambridge, Mass.: MIT Press, 1992.

Boyarin, Jonathan, ed. *The Ethnography of Reading*. Berkeley and Los Angeles: University of California Press, 1992.

Brendin, Jean-Denis. *The Affair: The Case of Alfred Dreyfus*. Trans. Jeffrey Mehlman. New York: George Braziller, 1986.

Brodhead, Richard H. *Cultures of Letters: Scenes of Reading and Writing in Nineteenth-Century America*. Chicago: University of Chicago Press, 1993.

Brown, David Wolfe. "The American Propagandism of Phonography." *The National Stenographer* 7-9 (1893): 283-91.

Browning, Robert Pack, Michael B. Frank, and Lin Salamo, eds. *Mark Twain's Notebooks and Journals*. Vol. 3. Berkeley and Los Angeles: University of California Press, 1979.

Brylawski, E. Fulton, and Abe Goldman, eds. *Legislative History of the 1909 Copyright Act*. 6 vols. South Hackensack, N.J.: Fred B. Rotheman, 1976.

Bull, Lucy C. "Being a Typewriter." *Atlantic Monthly* 76 (Dec. 1895): 822-31.

Burkett, Eva Mae. *A Study of American Dictionaries of the English Language before 1861*. Nashville, Tenn.: George Peabody College, 1936.

Campbell-Kelly, Martin, and William Aspray. *Computer: A History of the Information Machine*. New York: Basic Books, 1996.

Certeau, Michel de. *The Practice of Everyday Life*. Trans. Steven F. Rendall. Berkeley and Los Angeles: University of California Press, 1984.

Chandler, Alfred D., Jr. *The Visible Hand: The Managerial Revolution in American Business*. Cambridge: Harvard University Press, 1977.

Chartier, Roger. "Texts, Printing, Readings." In *The New Cultural History*, ed. Lynn Hunt, 154-75. Berkeley and Los Angeles: University of California Press, 1989.

Cheape, Charles W. *Moving the Masses: Urban Public Transit in New York, Boston, and Philadelphia, 1880-1912*. Cambridge: Harvard University Press, 1980.

Christy, Craig T. *Uniformitarianism in Linguistics*. Amsterdam Studies in the Theory and History of Linguistic Science, Series III, 31. Amsterdam: John Benjamins, 1983.

Cipolla, Carlo M. *Clocks and Culture, 1300-1700*. 1967. Reprint, New York: W. W. Norton, 1977.

Clayton, Jay. "The Voice in the Machine: Hazlitt, Hardy, James." In

Language Machines: Technologies of Literary and Cultural Production, ed. Jeffrey Masten, Peter Stallybrass, and Nancy Vickers, 209–32. New York: Routledge, 1997.

Clifford, James, and George E. Marcus, eds. *Writing Culture: The Poetics and Politics of Ethnography*. Berkeley and Los Angeles: University of California Press, 1986.

Cohen, Daniel A. *Pillars of Salt, Monuments of Grace: New England Crime Literature and the Origins of American Popular Culture*. New York: Oxford University Press, 1993.

Cohen, Lizabeth. *Making a New Deal: Industrial Workers in Chicago, 1919–1939*. New York: Cambridge University Press, 1990.

Collins, H. M. *Changing Order: Replication and Induction in Scientific Practice*. London: Sage Publications, 1985.

Cooper, Carolyn C. "Social Construction of Invention through Patent Management: Thomas Blanchard's Woodworking Machinery." *Technology and Culture* 32 (1991): 960–98.

Cooper, Jill E. "Intermediaries and Invention: Business Agents and the Edison Electric Pen and Duplicating Press." *Business and Economic History* 25 (1996): 130–42.

Coopersmith, Jonathan. *The Electrification of Russia, 1880–1926*. Ithaca, N.Y.: Cornell University Press, 1992.

Cortázar, Julio. *Around the Day in Eighty Worlds*. Trans. Thomas Christensen. San Francisco: North Point Press, 1986.

Cracas, Ronald. "Judge Learned Hand and the Law of Copyright." *American Society of Composers, Authors, and Publishers Copyright Law Symposium* 7 (1956): 55–90.

Crane, Hart. *The Bridge*. 1930. Reprint, New York: Liveright, 1992.

Crary, Jonathan. *Techniques of the Observer: On Vision and Modernity in the Nineteenth Century*. Cambridge, Mass.: MIT Press, 1990.

Current, Richard N. *The Typewriter and the Men Who Made It*. Urbana: University of Illinois Press, 1954.

Cutcliffe, Steven H., and Robert C. Post, eds. *In Context: History and the History of Technology, Essays in Honor of Melvin Kranzberg*. Bethlehem, Penn.: Lehigh University Press, 1989.

Czitrom, Daniel J. *Media and the American Mind from Morse to McLuhan*. Chapel Hill: University of North Carolina Press, 1982.

David, Paul. "Clio and the Economics of QWERTY." *American Economic Review. Papers and Proceedings* 75 (1985): 332–37.

———. "Heroes, Herds, and Hysteresis in Technological History: Thomas Edison and the 'Battle of the Systems' Reconsidered." *Industrial and Corporate Change* 1 (1992): 129–80.

Davidson, Cathy N. "The Life and Times of Charlotte Temple, The Biography of a Book." In *Reading in America: Literature and Social History*, ed. Cathy Davidson, 157–79. Baltimore: Johns Hopkins University Press, 1989.

Davies, Margery W. *Woman's Place Is at the Typewriter: Office Work and Office Workers, 1870–1930*. Philadelphia: Temple University Press, 1982.

Defendant's Testimony in the Case of A. J. Graham v. Benn Pitman. 1865. Reprint, Cincinnati, Ohio: Phonographic Institute, n.d.

Dégh, Linda. *American Folklore and the Mass Media*. Bloomington: Indiana University Press, 1994.

Democratic Souvenirs: An Historical Anthology of 19th-Century American Music. Ed. Richard Jackson. New York: C. F. Peters Corporation for the New York Public Library, 1988.

Derrida, Jacques. "Freud and the Scene of Writing." Trans. Jeffrey Mehlman. *Yale French Studies* 48 (1972): 73–117.

Dethlefson, Ronald. *Edison Blue Amberol Recordings, 1912–1914*. Brooklyn: APM Press, 1980.

Dickson, W. K. L., and Antonia Dickson. *History of the Kinetograph, Kinetoscope, and Kinetophonograph.* 1895. Reprint, New York: Arno Press, 1970.

Doheny-Farina, Stephen. *Rhetoric, Innovation, Technology: Case Studies of Technical Communication in Technology Transfers.* Cambridge, Mass.: MIT Press, 1992.

Dorman, James H. "Shaping the Popular Image of Post-Reconstruction American Blacks: The 'Coon Song' Phenomenon of the Guilding Age." *American Quarterly* 40 (1988): 450–71.

Douglas, Susan J. *Inventing American Broadcasting, 1899–1922.* Baltimore: Johns Hopkins University Press, 1987.

Dowling, Linda. *Language and Decadence in the Victorian Fin de Siècle.* Princeton, N.J.: Princeton University Press, 1986.

Doyle, Arthur Conan. "A Case of Identity." *Strand* 2 (1891): 248–59.

Dreiser, Theodore. "Concerning the Author of these Songs." In *The Songs of Paul Dresser*, by Paul Dresser, v–x. New York: Boni and Liveright, 1927.

Dresser, Paul. *The Songs of Paul Dresser.* New York: Boni and Liveright, 1927.

Duguid, Paul "Material Matters: The Past and Futerology of the Book." In *The Future of the Book*, ed. Geoffrey Nunberg, 63–102. Berkeley and Los Angeles: University of California Press, 1996.

DuMaurier, George. *Trilby*. New York: Dutton, 1931.

Dyer, Frank, and Thomas Commerford Martin. *Edison: His Life and Inventions.* 2 vols. New York: Harper and Bros., 1910.

Dyer, Richard. *Stars.* London: British Film Institute, 1979.

Eco, Umberto. *The Search for a Perfect Language.* Trans. James Fentress. Oxford: Blackwell, 1997.

Edel, Leon, and Lyall H. Powers, eds. *The Complete Notebooks of Henry James.* New York: Oxford University Press, 1987.

Edelman, Bernard. *Ownership of the Image: Elements for a Marxist Theory of Law.* Trans. Elizabeth Kingdom. London: Routledge and Keagan Paul, 1979.

Edison, Thomas A. *The Papers of Thomas A. Edison.* Ed. Reese V. Jenkins et al. 4 vols to date. Baltimore: Johns Hopkins University Press, 1989–.

———. "The Phonograph and Its Future." *North American Review* 126 (1878): 527–36.

———. *Thomas A. Edison Papers: A Selective Microfilm Edition.* Ed. Thomas E. Jeffrey et al. 4 parts to date. Bethesda, Md.: University Publications of America, 1985–.

Eisenstein, Elizabeth L. *The Printing Press as an Agent of Change: Communications and Cultural Transformations in Early-Modern Europe.* 2 vols. Cambridge: Cambridge University Press, 1979.

Ellis, Alexander J. *On Early English Pronunciation with Especial Reference to Shakspere and Chaucer, Containing an Investigation of the Correspondence of Writing with Speech in England from the Anglosaxon Period to the Present Day, Preceded by a Systematic Notation of All Spoken Sounds by Means of Ordinary Printing Types. Part I.* London: Philological Society, 1869.

Ezell, Margaret J. M., and Katherine O'Brien O'Keeffe, eds. *Cultural Artifacts and the Production of Meaning: The Page, the Image, and the Body.* Ann Arbor: University of Michigan Press, 1994.

Fabian, Johannes. "Keep Listening." In *The Ethnography of Reading,* ed. Jonathan Boyarin, 80–97. Berkeley and Los Angeles: University of California Press, 1992.

———. *Time and the Other: How Anthropology Makes Its Object.* New York: Columbia University Press, 1983.

Feather, John. *The Information Society: A Study of Continuity and Change.* London: Library Association Publishing, 1994.

Febvre, Lucien, and Henri-Jean Martin. *The Coming of the Book: The Impact of Printing, 1450–1800.* 1976. Reprint, ed. Geoffrey Nowell-Smith and David Wooton; trans. David Gerard, London: Verso, 1984.

Ferguson, Eugene S. *Engineering and the Mind's Eye.* Cambridge, Mass.: MIT Press, 1992.

———. "The Mind's Eye: Nonverbal Thought in Technology." *Science* 197 (1977): 827–36.

Fessenden, Thomas Green. *The Register of Arts, Or a Compendious View of Some of the Most Useful Modern Discoveries and Inventions.* Philadelphia: n.p., 1808.

Fischer, Claude S. "'Touch Someone': The Telephone Industry Discovers Sociability." In *Technology and Choice: Readings from Technology and Culture*, ed. Marcel C. LaFollette and Jeffrey K. Stine, 89–116. Chicago: University of Chicago Press, 1991.

Foucault, Michel. *Madness and Civilization: A History of Insanity in the Age of Reason.* Trans. Richard Howard. New York: Vintage Books, 1973.

Frank, Felicia Miller. *The Mechanical Song: Women, Voice, and the Artificial in Nineteenth-Century French Narrative.* Stanford: Stanford University Press, 1995.

Freud, Sigmund. *The Psychopathology of Everyday Life.* Ed. and trans. James Strachey. New York: W. W. Norton, 1989.

Frow, George L. *The Edison Disc Phonographs and the Diamond Discs.* Kent, Eng.: George L. Frow, 1982.

Frow, George L., and Albert F. Sefl. *The Edison Cylinder Phonographs.* Kent, Eng.: George L. Frow, 1985.

Gaines, Jane M. *Contested Culture: The Image, The Voice, and the Law.* Chapel Hill: University of North Carolina Press, 1991.

Garbit, Frederick F. *The Phonograph and Its Inventor, Thomas Alvah Edison.* Boston: n.p., 1878.

Garland, Hamlin. *Forty Years of Psychic Research: A Plain Narrative of Fact.* New York: MacMillan Company, 1936.

———. *A Son of the Middle Border.* 1917. Reprint, New York: MacMillan, 1930.

Geertz, Clifford. "Local Knowledge: Fact and Law in Comparative Perspective." In *Local Knowledge: Further Essays in Interpretive Anthropology*, by Clifford Geertz, 167–234. New York: Basic Books, 1983.

Gelatt, Roland. *The Fabulous Phonograph, From Tin Foil to High Fidelity.* Philadelphia: Lippincott, 1955.

Gilroy, Paul. *The Black Atlantic: Modernity and Double Consciousness.* Cambridge: Harvard University Press, 1993.

Ginzburg, Carlo. *The Cheese and the Worms: The Cosmos of a Sixteenth-Century Miller.* Trans. John and Anne Tedeschi. Baltimore: Johns Hopkins University Press, 1980.

Gitelman, Lisa. "The World Recounted: Science and Narrative in Early

Nineteenth-Century Exploration Accounts." Ph.D. diss., Department of English and Comparative Literature, Columbia University, 1991.

Goldberg, Jonathan. *Writing Matter: From the Hands of the English Renaissance.* Stanford: Stanford University Press, 1990.

Goody, Jack. *The Interface Between the Written and the Oral.* Cambridge: Cambridge University Press, 1987.

———. *The Logic of Writing and the Organization of Society.* Cambridge: Cambridge University Press, 1986.

Gorman, Michael E., and W. Bernard Carlson. "Interpreting Invention as a Cognitive Process: The Case of Alexander Graham Bell, Thomas Edison, and the Telephone." *Science, Technology, and Human Values* 15 (1990): 131–64.

Graff, Gerald. *Professing Literature: An Institutional History.* Chicago: University of Chicago Press, 1987.

Graham, Andrew Jackson. *Exhibits of the State of the Phonographic Art with Reference to the Copyright Case of Graham v. Pitman.* New York: A. J. Graham, 1864.

———. *The Hand-Book of Standard or American Phonography.* 5 parts. New York: A. J. Graham, 1886.

Grimsted, David. *Melodrama Unveiled: American Theatre and Culture, 1800–1850.* Chicago: University of Chicago Press, 1968.

Grint, Keith, and Steve Woolgar. *The Machine at Work: Technology, Work and Organization.* Cambridge, Eng.: Polity Press, 1997.

Guest, Ann Hutchinson. *Choreo-Graphics: A Comparison of Dance Notation Systems from the Fifteenth Century to the Present Day.* New York: Gordon and Breech, 1989.

Gumbrecht, Hans, and K. Ludwig Pfeiffer, eds. *Materialities of Communication.* Trans. William Whobrey. Stanford: Stanford University Press, 1994.

Hagen, Hugo von. *Reading Character from Handwriting: A Hand-Book of Graphology for Experts, Students, and Laymen.* New York: Graphology Publishing Co., 1902.

Hakutani, Yoshinobu, ed. *Selected Magazine Articles of Theodore Dreiser: Life and Art in the American 1890s.* 2 vols. Rutherford, N.J.: Farleigh Dickenson University Press, 1987.

Harraway, Donna J. *Simians, Cyborgs, and Women: The Reinvention of Nature.* New York: Routledge, 1991.

Harris, Neil. *Cultural Excursions: Marketing Appetites and Cultural Tastes in Modern America.* Chicago: University of Chicago Press, 1990.

Harvey, David. *The Condition of Postmodernity: An Enquiry into the*

Origins of Cultural Change. Cambridge, Mass.: Basil Blackwell, 1990.

Havelock, Eric A. *Preface to Plato.* Cambridge: Harvard University Press, 1963.

Hoke, Donald R. *Ingenious Yankees: The Rise of the American System of Manufactures in the Private Sector.* New York: Columbia University Press, 1990.

Hollow, Robert C. "Concordance of Phonetic Alphabets." *Kroeber Anthropological Society Papers,* no. 43 (1970): 42–50.

Holzer, Harold, ed. *Dear Mr. Lincoln: Letters to the President.* Reading, Mass.: Addison-Wesley, 1993.

Hounshell, David. "Elisha Gray and the Telephone: On the Disadvantages of Being an Expert." *Technology and Culture* 16 (1975): 135–61.

Howells, William Dean. *The Undiscovered Country.* 1880. Reprint, Boston: Houghton, Mifflin, and Co., 1893.

Hower, Ralph M. *History of Macy's New York, 1858–1919: Chapters in the Evolution of the Department Store.* Cambridge: Harvard University Press, 1946.

Hubert, Philip G., Jr. "The New Talking-Machines." *Atlantic Monthly* 63 (1889): 256–61.

Humboldt, Wilhelm von, *On Language: The Diversity of Human Language-Structure and Its Influence on the Mental Development of Mankind.* Trans. Peter Heath. Cambridge: Cambridge University Press, 1988.

Hunt, Stocker. *Ouija: The Most Dangerous Game.* New York: Harper and Row, 1985.

Hunter, Ian M. "Lengthy Verbatim Recall: The Role of Text." *Progress in the Psychology of Language.* Vol. 1. Ed. Andrew W. Ellis. Hillsdale, N.J.: Lawrence Erlbaum, 1985.

Irwin, John T. *American Hieroglyphics: The Symbol of the Egyptian Hieroglyphics in the American Renaissance.* New Haven, Conn.: Yale University Press, 1980.

Isaacs, Ernest. "The Fox Sisters and American Spiritualism." In *The Occult in America: New Historical Perspectives,* ed. Howard Kerr and Charles L. Crow, 79–110. Urbana: University of Illinois Press, 1983.

Israel, Paul. *Edison: A Life of Invention.* New York: John Wiley and Sons, 1998.

Israel, Paul, and Robert Rosenberg. "Patent Office Records as a Historical Source: The Case of Thomas Edison." *Technology and Culture* 32 (1991): 1094–101.

James, William. "A Case of Automatic Drawing." *The Popular Science Monthly* 64 (Jan. 1904): 195–201.

Jameson, Fredric. *The Political Unconscious: Narrative as a Socially Symbolic Act.* Ithaca, N.Y.: Cornell University Press, 1981.

Jenkins, Reese V. "Elements of Style: Continuities in Edison's Thinking." *Annals of the New York Academy of Sciences* 424 (1984): 149–62.

John, Richard R. *Spreading the News: The American Postal System from Franklin to Morse.* Cambridge, Mass.: Harvard University Press, 1995.

Johnson, James H. *Listening in Paris: A Cultural History.* Berkeley and Los Angeles: University of California Press, 1995.

Kaplan, Benjamin. *An Unhurried View of Copyright.* New York: Columbia University Press, 1967.

Kasson, John F. *Amusing the Million: Coney Island at the Turn of the Century.* New York: Hill and Wang, 1978.

Kenner, Hugh. *The Mechanic Muse.* New York: Oxford University Press, 1987.

Kern, Stephen. *The Culture of Time and Space, 1880–1918.* Cambridge: Harvard University Press, 1983.

Kerr, Howard. *Mediums, and Spirit-Rappers, and Roaring Radicals: Spiritualism in American Literature, 1850–1900.* Urbana: University of Illinois Press, 1972.

Kittler, Friedrich A. *Discourse Networks 1800/1900.* Trans. Michael Metteer, with Chris Cullens. Stanford: Stanford University Press, 1990.

———. *Gramophone, Film, Typewriter.* Trans. Geoffrey Winthrop-Young and Michael Wutz. Stanford, Calif.: Stanford University Press, 1999.

Knoper, Randall. *Acting Naturally: Mark Twain and the Culture of Performance.* Berkeley and Los Angeles: University of California Press, 1995.

Koenigsberg, Allen. *Edison Cylinder Records, 1889–1912.* 2d ed. Brooklyn: APM Press, 1987.

———. *The Patent History of the Phonograph, 1877–1912.* 2d ed. Brooklyn: APM Press, 1991.

Koutstaal, Wilma. "Skirting the Abyss: A History of Experimental Explorations of Automatic Writing in Psychology." *Journal of the History of the Behavioral Sciences* 28 (Jan. 1992): 5–27.

Kwoleck-Folland, Angel. *Engendering Business: Men and Women in the Corporate Office, 1870–1930.* Baltimore: Johns Hopkins University Press, 1994.

Laing, Dave. "A Voice Without a Face: Popular Music and the Phonograph in the 1890s." *Popular Music* 10 (1991): 1–9.

Landow, George P. *Hypertext: The Convergence of Contemporary Critical Theory and Technology*. Baltimore: Johns Hopkins University Press, 1993.

———. "Twenty Minutes into the Future, Or How Are We Moving Beyond the Book?" In *The Future of the Book*, ed. Geoffrey Nunberg, 209–73. Berkeley and Los Angeles: University of California Press, 1996.

Lanham, Richard A. *The Electronic Word: Democracy, Technology, and the Arts*. Chicago: University of Chicago Press, 1993.

Lathrop, George Parsons. "Talks with Edison." *Harper's New Monthly Magazine* (Feb. 1890): 425–35.

Latour, Bruno. "Drawing Things Together." In *Representation in Scientific Practice*, ed. Michael Lynch and Steve Woolgar, 19–68. Cambridge, Mass.: MIT Press, 1990.

———. *Science in Action: How to Follow Scientists and Engineers through Society*. Cambridge: Harvard University Press, 1987.

Latour, Bruno, and Steve Woolgar. *Laboratory Life: The Construction of Scientific Facts*. 1979. Reprint, Princeton, N.J.: Princeton University Press, 1986.

Layton, Edwin T., Jr. "Mirror-Image Twins: The Communities of Science and Technology in 19th-Century America." *Technology and Culture* 12 (1971): 562–80.

———. "Technology as Knowledge." *Technology and Culture* 15 (1974): 31–41.

Lears, Jackson. *Fables of Abundance: A Cultural History of Advertising in America*. New York: Basic Books, 1994.

Leary, David E. "Telling Likely Stories: The Rhetoric of the New Psychology, 1880–1920." *Journal of the History of the Behavioral Sciences* 23 (Oct. 1987): 315–31.

Leonard, Thomas C. *The Power of the Press: The Birth of American Political Reporting*. New York: Oxford University Press, 1986.

Leppert, Richard. *The Sight of Sound: Music, Representation, and the History of the Body*. Berkeley and Los Angeles: University of California Press, 1993.

Levin, Thomas Y. "For the Record: Adorno on Music in the Age of Its Technological Reproducibility." *October* 55 (1990): 23–66.

Levine, Lawrence W. "The Folklore of Industrial Society: Popular Culture and Its Audiences." *American Historical Review* (Dec. 1992): 1369–99.

———. *Highbrow/Lowbrow: The Emergence of Cultural Hierarchy in America*. Cambridge: Harvard University Press, 1988.

Lewis, George H. "The Meanings in the Music and the Music's in Me: Popular Music as Symbolic Communication." *Theory, Culture and Society* 1 (1983): 133–41.

Liebowitz, S. J., and Stephen E. Margolis. "The Fable of the Keys." *Journal of Law and Economics* 33 (1990): 1–26.

Looby, Christopher. *Voicing America: Language, Literary Form, and the Origins of the United States.* Chicago: University of Chicago Press, 1996.

Lott, Eric. *Love and Theft: Blackface Minstrelsy and the American Working Class:* New York: Oxford University Press, 1993.

Loughney, Patrick George. "A Descriptive Analysis of the Library of Congress Paper Print Collection and Related Copyright Materials." Ph.D. diss., Department of American Civilization, George Washington University, 1988.

Lovett, Charles C. "Charles Dodgson and His Hammond Type Writer." *Jabberwocky: The Journal of the Lewis Carrol Society* 19 (1990): 3–9.

Macey, Samuel L. *The Dynamics of Progress: Time, Method, and Measure.* Athens: University of Georgia Press, 1989.

MacLeod, Christine. "The Paradoxes of Patenting: Invention and Its Diffusion in 18th- and 19th-Century Britain, France, and America." *Technology and Culture* 32 (1991): 885–910.

Macomber, William. *The Fixed Law of Patents.* Boston: Little, Brown, 1909.

Mandell, Richard D. *Paris 1900: The Great World's Fair.* Toronto: University of Toronto Press, 1976.

Mares, George Carl. *The History of the Typewriter: Being an Illustrated Account of the Origin, Rise, and Development of the Writing Machine.* London: Guilbert Pitman, 1909.

Martin, Henri-Jean. *The History and Power of Writing.* Trans. Lydia G. Cochrane. Chicago: University of Chicago Press, 1994.

Marvin, Carolyn. *When Old Technologies Were New: Thinking About Electric Communication in the Late Nineteenth Century.* New York: Oxford University Press, 1988.

Marx, Leo. *The Machine in the Garden: Technology and the Pastoral Ideal in America.* New York: Oxford University Press, 1964.

Masi, Frank T., ed. *The Typewriter Legend.* Secaucus, N.J.: Matsushita Electric Corporation of America, 1985.

Masten, Jeffrey, Peter Stallybrass and Nancy Vickers, eds. *Language Machines: Technologies of Literary and Cultural Production.* New York: Routledge, 1997.

McCray, Florine Thayer. "The Type-Writer as an Aid to Fluent
Composition." *The Writer* 3 (1889): 114–15.

McDermut, W. E. "Annual Convention, National Shorthand Reporters'
Association." *Life and Labor* 10 (1911): 334–35.

McElvaine, Robert S., ed. *Down and Out in the Great Depression:
Letters from the Forgotten Man*. Chapel Hill: University of North
Carolina Press, 1983.

McGann, Jerome J. *The Textual Condition*. Princeton, N.J.: Princeton
University Press, 1991.

McGill, Meredith L. "The Matter of the Text: Commerce, Print Culture,
and the Authority of the State in American Copyright Law." *American
Literary History* 9 (1997): 21–59.

McKee, Charles M. *The New Rapid: A Light-Line Connective-Vowel
System of Shorthand*. 10th ed. Buffalo, N.Y.: McKee, 1896.

McKenzie, D. F. *Bibliography and the Sociology of Texts*. London: British
Library, 1986.

McKenzie, Donald. *Knowing Machines: Essays on Technical Change*.
Cambridge, Mass.: MIT Press, 1996.

McLuhan, Marshall. *Understanding Media: The Extensions of Man*.
New York: McGraw-Hill, 1964.

Meadowcroft, William H. "The Story of the Phonograph." *St. Nicholas*
49 (May 1922): 697–99.

Meikle, Jeffrey L. *American Plastic, A Cultural History*. New Brunswick,
N.J.: Rutgers University Press, 1995.

Mencken, H. L. *The American Language: An Inquiry into the
Development of English in the United States*. 4th ed. New York:
Alfred A. Knopf, 1937.

Middleton, Richard. *Studying Popular Music*. Milton Keyenes, Eng.:
Open University Press, 1990.

Millard, Andre. *Edison and the Business of Innovation*. Baltimore:
Johns Hopkins University Press, 1990.

Miller, Perry, ed. *The Legal Mind in America, From Independence to the
Civil War*. Ithaca, N.Y.: Cornell University Press, 1962.

Milosi, Martin V. *Thomas A. Edison and the Modernization of America*.
Glenview, Ill.: Scott, Foresman, Little, Brown Higher Education, 1990.

Mumford, Lewis. *Technics and Civilization*. New York: Harcourt, 1934.

Munson, James E. *The Complete Phonographer and Reporter's Guide: An
Inductive Exposition of Phonography*. New York: Harper, 1884.

Murphy, Garner, and Robert O. Ballou, eds. *William James on Psychical
Research*. Clifton, N.J.: August M. Kelly, 1973.

Musser, Charles. *Before the Nickelodeon: Edwin S. Porter and the Edison*

Manufacturing Company. Berkeley and Los Angeles: University of California Press, 1991.

———. *The Emergence of Cinema: The American Screen to 1907.* New York: Charles Scribner's Sons, 1990.

Myers, Greg. "From Discovery to Invention: The Writing and Rewriting of Two Patents." *Social Studies of Science* 25 (1995): 57–105.

Nasaw, David. *Going Out: The Rise and Fall of Public Amusements.* New York: Basic Books, 1993.

Nord, David Paul. "Reading the Newspaper: Strategies and Politics of Reader Response, Chicago 1912–1917." *Journal of Communication* 45 (1995): 66–93.

Norman, Donald A. *The Psychology of Everyday Things.* New York: Basic Books, 1988.

Nunberg, Geoffrey, "Farewell to the Information Age." In *The Future of the Book*, ed. Geoffrey Nunberg, 103–38. Berkeley and Los Angeles: University of California Press, 1996.

Nye, David E. *American Technological Sublime.* Cambridge, Mass.: MIT Press, 1994.

———. *Electrifying America: Social Meanings of a New Technology, 1880–1940.* Cambridge, Mass.: MIT Press, 1990.

O'Brien, Robert Lincoln. "Machinery and English Style." *Atlantic Monthly* 94 (1904): 464–72.

Olson, David R. *The World on Paper: The Conceptual and Cognitive Implications of Writing and Reading.* Cambridge: Cambridge University Press, 1994.

O'Malley, Michael. "Specie and Species: Race and the Money Question in Nineteenth-Century America." *American Historical Review* 99 (1994): 369–95.

Ong, Walter J. *Orality and Literacy: The Technologizing of the Word.* New York: Methuen, 1982.

Orvell, Miles. *The Real Thing: Imitation and Authenticity in American Culture, 1880–1940.* Chapel Hill: University of North Carolina Press, 1989.

Pacey, Arnold. *The Maze of Ingenuity: Ideas and Idealism in the Development of Technology.* New York: Holmes and Meier, 1975.

Painter, Nell Irvin. "Thinking About the Languages of Money and Race: A Response to Michael O'Malley, 'Specie and Species.'" *American Historical Review* 99 (1994): 396–404.

Panchasi, Roxanne. "Graphology and the Science of Individual Identity in Modern France." *Configurations* (1996): 1–31.

Paulson, William R. *The Noise of Culture: Literary Texts in a World of Information.* Ithaca, N.Y.: Cornell University Press, 1988.

Peiss, Kathy. *Cheap Amusements: Working Women and Leisure in Turn-of-the-Century New York.* Philadelphia: Temple University Press, 1986.

Peitz, William. "The Phonograph in Africa: International Phonocentrism from Stanley to Sarnoff." In *Post-Structuralism and the Question of History,* ed. Derek Attridge, Geoff Bennington, and Robert Young. Cambridge: Cambridge University Press, 1987.

Petroski, Henry. *The Pencil: A History of Design and Circumstance.* New York: Alfred A. Knopf, 1990.

Pitman, Benn. *The Manual of Phonography.* Cincinnati, Ohio: Phonographic Institute, 1860.

Pitman, Sir James, and John St. John. *Alphabets and Reading: The Initial Teaching Alphabet.* New York: Pitman Publishing, 1969.

Porter, Theodore M. *Trust in Numbers: The Pursuit of Objectivity in Science and Public Life.* Princeton, N.J.: Princeton University Press, 1995.

Pratt, Mary Louise. "Fieldwork in Common Places." In *Writing Culture: The Poetics and Politics of Ethnography,* ed. James Clifford and George E. Marcus, 27–50. Berkeley and Los Angeles: University of California Press, 1986.

Preece, William H., and Augustus Stroh. "Studies in Acoustics." *Proceedings of the Royal Society of London* 28 (Feb. 1879): 358–66.

Proceedings of the First Annual Convention of Local Phonograph Companies. 1890. Reprint, Nashville, Tenn.: Country Music Foundation Press, 1974.

Proceedings of the New York State Law Stenographers Association 4 (1879): n.p.

Purcell, L. Edward. "Trilby and Trilby-Mania, the Beginning of the Bestseller System." *Journal of Popular Culture* 11 (1977): 62–76.

Purdy, Strother B. "Technopoetics: Seeing What Literature Has to Do with the Machine." *Critical Inquiry* 11 (1984): 130–40.

Rama, Angel. *The Lettered City.* Ed. and trans. John Charles Chasteen. Durham, N.C.: Duke University Press, 1996.

Read, Oliver, and Walter L. Welch. *Tin Foil to Stereo: Evolution of the Phonograph.* 2d ed. Indianapolis, Ind.: Howard W. Sams and Co., 1976.

Redmond, Donald A. *Sherlock Holmes among the Pirates: Copyright and Conan Doyle in America, 1890–1930.* New York: Greenwood Press, 1990.

Reiser, Stanley Joel. *Medicine and the Reign of Technology.* Cambridge: Cambridge University Press, 1978.

Riis, Thomas L. *Just Before Jazz: Black Musical Theatre in New York, 1890–1915.* Washington, D.C.: Smithsonian Institution, 1989.

Rosenberg, Richard S. *The Social Impact of Computers.* 1986. Reprint, San Diego, Cal.: Academic Press, 1992.

Ruhleder, Karen. "Reconstructing Artifacts, Reconstructing Work: From Textual Edition to On-Line Databank." *Science, Technology, and Human Values* 20 (1995): 39–64.

Sanjek, Russell. *American Popular Music and Its Business: The First Four Hundred Years.* Vol. 2.: *From 1790 to 1909.* New York: Oxford University Press, 1988.

Schaffer, Simon. "Self Evidence." *Critical Inquiry* 18 (1992). Reprinted in *Questions of Evidence: Proof, Practice, and Persuasion across the Disciplines*, ed. James Chandler et al., 56–91, Chicago: University of Chicago Press, 1994.

Schudson, Michael. *Origins of the Ideal of Objectivity in the Professions: Studies in the History of American Journalism and American Law, 1830–1940.* New York: Garland, 1990.

Schwartzman, Arnold. *Phono-Graphics: The Visual Paraphrenalia of the Talking Machine.* Photography by Garry Brod. San Francisco: Chronicle Books, 1993.

Scovil, W. E. *A Short-Hand, Legible as the Plainest Writing, and Requiring No Teacher but the Book.* 18th ed. New York: n.p., 1873.

Scranton, Philip. "Determinism and Indeterminacy in the History of Technology." In *Does Technology Drive History: The Dilemma of Technological Determinism*, ed. Merritt Roe Smith and Leo Marx, 143–68. Cambridge, Mass.: MIT Press, 1994.

Scribner, Sylvia, and Michael Cole. *The Psychology of Literacy.* Cambridge: Harvard University Press, 1981.

Sebastian, Thomas. "Technology Romanticized: Friedrich Kittler's *Discourse Networks 1800/1900.*" *MLN* 105 (1990): 583–95.

Segal, Howard P. *Technological Utopianism in American Culture.* Chicago: University of Chicago Press, 1985.

Shapin, Steven. "Pump and Circumstance: Robert Boyle's Literary Technology." *Social Studies of Science* 14 (1984): 481–520.

———. *A Social History of Truth.* Chicago: University of Chicago Press, 1994.

Shapin, Steven, and Simon Schaffer. *Leviathan and the Air Pump: Hobbes, Boyle, and the Experimental Life.* Princeton, N.J.: Princeton University Press, 1985.

Shurkin, Joel. *Engines of the Mind: The Evolution of the Computer from Mainframes to Microprocessors.* 1984. Reprint, New York: W. W. Norton, 1996.

Siefert, Marsha. "Aesthetics, Technology, and the Capitalization of Culture: How the Talking Machine Became a Musical Instrument." *Science in Context* 8 (1995): 417–49.

Simpson, David. *The Politics of American English, 1776–1850.* New York: Oxford University Press, 1986.

Smith, Henry Nash, and William M. Gibson, eds. *Mark Twain—Howells Letters: The Correspondence of Samuel L. Clemens and William Dean Howells, 1872–1910.* 2 vols. Cambridge: Harvard University Press, 1960.

Smith, Merritt Roe, and Leo Marx, eds. *Does Technology Drive History? The Dilemma of Technological Determinism.* Cambridge, Mass.: MIT Press, 1994.

Snyder, Robert W. *The Voice of the City: Vaudeville and Popular Culture in New York.* New York: Oxford University Press, 1989.

Staudenmaier, John M. *Technology's Storytellers: Reweaving the Human Fabric.* Cambridge, Mass.: MIT Press, 1985.

Stein, Gertrude. "Cultivated Motor Automatisms: A Study of Character in Its Relation to Attention." *Psychological Review* 5 (1898): 295–306.

Stein, Gertrude, and Louis M. Solomons. "Normal Motor Automatisms." *Psychological Review* 3 (1896): 492–512.

Stewart, Susan. *Crimes of Writing: Problems in the Containment of Representation.* Durham, N.C.: Duke University Press, 1994.

———. *On Longing, Narratives of the Miniature, the Gigantic, the Souvenir, the Collection.* Durham, N.C.: Duke University Press, 1993.

Stingelin, Martin, "Comments on a Ball: Nietzche's Play on the Typewriter." In *Materialities of Communication,* ed. Hans Gumbrecht and K. Ludwig Pfeiffer; trans. William Whobrey, 70–82. Stanford: Stanford University Press, 1994.

Stoker, Bram. *Dracula.* 1897. Reprint, New York: Penguin Books, 1992.

Stokes, J. Lort. *Discoveries in Australia.* 2 vols. London: n.p., 1846.

Stowe, David W. "Passing through Air: Radio and Race Ventriloquism during the 1920s and 1930s." Paper presented at the American Studies Association, Pittsburgh, Penn., Nov. 10, 1995.

Strasser, Susan. *Satisfaction Guaranteed: The Making of the American Mass Market.* New York: Pantheon, 1989.

Street, Brian V. Introduction to *Cross-Cultural Approaches to Literacy,* ed. Brian Street. Cambridge: Cambridge University Press, 1993.

Strout, Cushing. *Making American Tradition: Visions and Revisions from Ben Franklin to Alice Walker.* New Brunswick, N.J.: Rutgers University Press, 1990.

Surrency, Erwin C. "Law Reports in the United States." *American Journal of Legal History* 25 (1981): 48–66.

Talbot, William Henry Fox. *The Pencil of Nature.* London: n.p., 1844.

Taussig, Michael. *Mimesis and Alterity: A Particular History of the Senses.* New York: Routledge, 1993.

Tenner, Edward. *Why Things Bite Back: Technology and the Revenge of Unintended Consequences.* New York: Alfred A. Knopf, 1996.

Thompson, Emily. "Machines, Music, and the Quest for Fidelity: Marketing the Edison Phonograph in America, 1877–1925." *The Musical Quarterly* 79 (1995): 131–71.

Tibbetts, Paul. "Representation and the Realist-Constructivist Controversy." In *Representation in Scientific Practice*, ed. Michael Lynch and Steve Woolgar, 69–84. Cambridge, Mass.: MIT Press, 1990.

Tinling, Marion. "Thomas Lloyd's Reports of the First Federal Congress." *William and Mary Quarterly*, 3d series, 18 (1961): 519–45.

Towndrow, Thomas. *A Complete Guide to the Art of Writing Short-Hand.* Boston: Perkins and Marvin, 1832.

Transactions of the First International Shorthand Congress. London: I. Pitman and Sons, 1888.

Turner, Patricia A. *Ceramic Uncles and Celluloid Mammies: Black Images and Their Influence on Culture.* New York: Anchor Books, 1994.

———. *I Heard It through the Grapevine: Rumor in African-American Culture.* Berkeley and Los Angeles: University of California Press, 1993.

Twain, Mark. "First Writing Machines." In *The $30,000 Bequest and Other Stories*, 224–28 of the *Complete Works of Mark Twain*, vol. 18. New York: Harper and Bros., 1906.

Underwood, B. F. "'Automatic' Writing." *Arena* 6 (1896): 74–83.

Vaughan, Floyd L. *The United States Patent System: Legal and Economic Conflicts in American Patent History.* 1956. Reprint, Westport, Conn.: Greenwood Press, 1972.

Venturi, Robert. *Learning from Las Vegas: The Forgotten Symbolism of Architectural Form.* Cambridge, Mass.: MIT Press, 1977.

Virilio, Paul. *The Vision Machine.* Trans. Julie Rose. Bloomington: University of Indiana Press and the British Film Institute, 1994.

Wachhorst, Wyn. *Thomas Alva Edison: An American Myth.* Cambridge, Mass.: MIT Press, 1981.

Wajcman, Judy. *Feminism Confronts Technology.* University Park: Pennsylvania State University Press, 1991.

Walker, Albert H. *Text-Book of The Patent Laws of the United States of America.* 4th ed. New York: Baker, Voorhis, 1904.

Warner, Michael. *The Letters of the Republic Publication and the Public Sphere in Eighteenth-Century America.* Cambridge: Harvard University Press, 1990.

Weber, Robert J. *Forks, Phonographs, and Hot Air Balloons: A Field Guide to Inventive Thinking.* New York: Oxford University Press, 1992.

Webster, Epinetus. *The Phonographic Teacher: Being an Inductive Exposition of Phonography.* New York: Fowler and Wells, 1852.

Wellbery, David E. Foreword to *Discourse Networks 1800/1900*, by Friedrich A. Kittler; trans. Michael Metteer with Chris Cullens, vii–xxxiii. Stanford: Stanford University Press, 1990.

Wicke, Jennifer. "Vampiric Typewriting: *Dracula* and Its Media." *ELH* 59 (1992): 467–93.

Wik, Reynold M. *Henry Ford and Grass-Roots America.* Ann Arbor: University of Michigan Press, 1972.

Wilde, Oscar. *Complete Works of Oscar Wilde.* New York: Harper and Row, 1989.

Wilkes, Charles. *Narrative of the United States Exploring Expedition.* 5 vols. 1845. Reprint, Upper Saddle River, N.J.: Gregg Press, 1970.

Winner, Langdon. "Do Artifacts Have Politics?" *Daedalus* 109 (1980): 121–36.

———. "Upon Opening the Black Box and Finding It Empty: Social Constructivism and the Philosophy of Technology." *Science, Technology, and Human Values* 18 (1993): 362–78.

Woodmansee, Martha, and Peter Jaszi, eds. *The Construction of Authorship: Textual Appropriation in Law and Literature.* Durham, N.C.: Duke University Press, 1994.

Woodward, A. L. "The Witness-Box and Its Occupants." *Proceedings of the New York State Stenographers' Association* 5 (1880): 33–36.

Yates, JoAnne. *Control through Communication: The Rise of System in American Management.* Baltimore: Johns Hopkins University Press, 1989.

Zboray, Ronald. *Fictive People: Antebellum Economic Development and the American Reading Public.* New York: Oxford University Press, 1993.

Index

In this index an "f" after a number indicates a separate reference on the next page, and an "ff" indicates separate references on the next two pages. A continuous discussion over two or more pages is indicated by a span of page numbers, e.g., "57–59." *Passim* is used for a cluster of references in close but not consecutive sequence.

Library of Congress Cataloging-in-Publication Data

Gitelman, Lisa.
    Scripts, grooves, and writing machines : representing technology
in the Edison era / Lisa Gitelman.
        p.   cm.
    Includes bibliographical references
    ISBN 0-8047-3270-1 (cl. : alk. paper). — ISBN 0-8047-3872-6
(pa. : alk. paper)
        1. Communication and technology—United States—History.
2. Literacy—Technological innovations—United States—History.
I. Title.
P96.T422U6343
302.2'0973—dc21   99-30422

Original printing 1999
Last figure below indicates year of this printing:
08   07   06   05   04   03   02   01   00   99

Designed by James P. Brommer
Typeset in 10/14 Sabon and Helvetica Black display

Printed and bound by CPI Group (UK) Ltd, Croydon, CR0 4YY

24/09/2024

14562869-0001